WARM HOUSE
COOL HOUSE

WARM HOUSE
COOL HOUSE

Money-Saving Guide to Energy Use in Your Home

The Editors
of Consumer Reports with
Monte Florman

Consumer Reports Books
A Division of Consumers Union
Yonkers, New York

Copyright © 1991 by
Consumers Union of United States, Inc.,
Yonkers, New York 10703
All rights reserved, including the right
of reproduction in whole or in part in any form.

Library of Congress Cataloging-in-Publication Data
Warm house, cool house : money-saving guide to energy use in your home
    / the editors of Consumer reports with Monte Florman.
        p.   cm. — (The Homeowner's library)
    Includes index.
    ISBN 0-89043-404-2 (pb)
    1. Dwellings—Energy conservation.   I. Florman, Monte.
II. Consumer reports.   III. Series.
TJ163.5.D86W36   1991
696—dc20   91-29374
                                                               CIP

Design by Tammy O'Bradovich

First Printing, November 1991
Manufactured in the United States of America

Warm House/Cool House is a Consumer Reports Book published by Consumers Union, the nonprofit organization that publishes *Consumer Reports,* the monthly magazine of test reports, product Ratings, and buying guidance. Established in 1936, Consumers Union is chartered under the Not-For-Profit Corporation Law of the State of New York.

The purposes of Consumers Union, as stated in its charter, are to provide consumers with information and counsel on consumer goods and services, to give information on all matters relating to the expenditure of the family income, and to initiate and to cooperate with individual and group efforts seeking to create and maintain decent living standards.

Contents

FOUR
Central Heating Systems

FIVE
Supplementary Heating

SIX
Water Heating

SEVEN
Keeping Cool

EIGHT
Saving Energy Dollars with Appliances and Lighting

NINE
Contracting Work: When Do-It-Yourself Won't Do

Introduction

Fundamentally, saving energy is saving money. This book is a compilation of practical information on how to save energy—and money—in your home. You may already have taken some energy-saving measures by buying storm windows, say, or by having your heating system tuned up. Even so, as fuel costs continue to rise, you may be considering a wide range of other steps to combat the cost of energy. If you have had a home energy audit, you should have a good idea of the specific energy-savers you need. If you haven't had an audit, the options may seem bewildering.

This book will help you decide which strategies are best suited to your needs. Some sections include worksheets that will aid you in estimating the savings that specific conservation measures will generate in your home.

Much of the material on energy saving has been prepared by using traditional methods of calculating energy consumption. In those cases where the standard methods have shortcomings, they have been modified. Still, the conventional methods frequently are the best way (sometimes the only way) available for calculating energy consumption and energy saving.

In discussing ways to save energy—and money—estimates of possible monetary saving have been based on the nationwide average costs for energy in 1990, as computed by the federal government:

- Fuel oil: 88 cents per gallon
- Natural gas: 56 cents per 100 cubic feet, which typically has an energy content of approximately one therm, or 100,000 Btu. (In this book, unless otherwise specified, *gas* means natural gas.)
- Electricity: 7.9 cents per kilowatt hour (kwh)

THE RATINGS

Individual brands and models are rated based on the estimated quality of the tested product samples. The Ratings order is derived from laboratory tests, controlled-use tests, and/or expert judgments. The Ratings offer comparative buying information that greatly increases the likelihood that you will receive value for your money. To take best advantage of the Ratings, first read the introduction preceding each chart, and then the notes and footnotes, in order to find out about the features, qualities, or deficiencies shared by products in the test group. The first sentence in the introduction to each Ratings chart provides the specific basis of the Ratings order.

PRICES

Where a listing of brands and models appears, you will find a notation of the month and year in which the Ratings appeared in *Consumer Reports.* Usually the prices listed are those published in the original report. Regardless of how current the prices are when you shop, remember to use the printed price only as a guide, since discounts are so widely available. It pays to shop around whenever possible, and buy from the dealer or source that offers the best price and satisfactory servicing arrangements.

MODEL CHANGES

Manufacturers commonly change products from time to time in an effort to stimulate sales, to keep up with or stay ahead of the competition, and/or to incorporate technological improvements. Unfortunately, dealer inventories cannot always keep pace with these changes, and older models are often still available long after newer ones have been introduced. When a dealer's carry-over inventory is large, older models may remain available for months or even years after they have been discontinued. As you negotiate a sale, you can often save money by being aware of what is current and what is not—keeping in mind that if you buy an older model, you may not always be getting a product that is on the cutting edge of the latest technology.

Even though a particular brand and model you select from the Ratings may be out of stock or superseded by a newer version, the information supplied here can be a great help in sorting out the products and their characteristics.

Most of the material in this book originally appeared in one form or another in *Consumer Reports.* Other sources are noted in the text. Everything has been reviewed and revised by Consumers Union's technical staff, by the author, and by the editors of Consumer Reports Books.

ONE

Your House Is a Complex Energy System

World events dramatize the vital role that energy conservation plays in our lives. With no decline in sight for the high price of energy, householders increasingly are seeking ways to cut back on energy use. There are dozens of steps you can consider for saving energy in your home. Some will yield a dramatic saving; others will have far more modest results; still others simply won't work in certain cases.

To know which energy-saving measures will be the most effective for you, it's best to understand some basic information about energy and its uses in the home.

HOW YOUR HOUSE USES ENERGY

In addition to its heating system, a house contains dozens of devices—light bulbs, kitchen appliances, television sets, radios, and more—that use energy in one way or another, ultimately releasing it in the form of heat.

To understand how energy is used in the home, we have to look beyond a specific task to see how that task relates to all other energy-consuming equipment in the home. In short, we have to think of the house as an "energy system."

As an example of how the system works, consider a hypothetical family at home in midafternoon on a cold, blustery day. The windows are shut and the furnace is running. Lights are on in one or two rooms, there is a load of clothes in the dryer, and a pie is baking in the oven. One member of the family is taking a shower while the others are watching television. The energy system is hard at work. Oil is being burned in the furnace to supply heat for the rooms; gas is being burned in the water heater so there is hot water for the shower. Electricity is providing heat to dry the clothes and bake the pie. Electricity also is providing the room light and running the television set.

While the furnace supplies much of the heat in our hypothetical house, it's not the only heat source. All the activities—showering, cooking, laundering, and so on—are contributing heat to the system. Even the people themselves

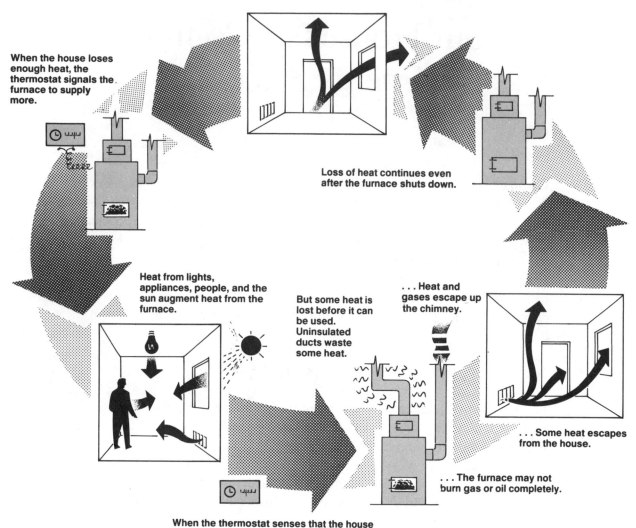

When the house loses enough heat, the thermostat signals the furnace to supply more.

Loss of heat continues even after the furnace shuts down.

Heat from lights, appliances, people, and the sun augment heat from the furnace.

But some heat is lost before it can be used. Uninsulated ducts waste some heat.

. . . Heat and gases escape up the chimney.

. . . Some heat escapes from the house.

. . . The furnace may not burn gas or oil completely.

When the thermostat senses that the house is warm enough, it signals the furnace to stop.

Figure 1.1 Heating cycle

are supplying some heat—as are the rays of the winter sun streaming through the windows.

At some point, the house will warm up sufficiently. (For practical purposes, that point is the thermostat setting selected by the family.) When the thermostat senses that the desired temperature has been reached, it signals the heating system to shut down until heat is needed again. The house will stay comfortable, but only for a while, because the heat inside the house will escape faster than all the other heat sources can supply replacement heat. One avenue of heat loss is *conduction*—the flow of heat through solid substances, such as ceilings, walls, floors, and windows. A second avenue is *air leakage*—the movement of heated air out through cracks around doors, windows, and other gaps in the house.

Some appliances can also contribute to air leakage. A kitchen exhaust fan, for example, will remove heated air as well as cooking odors. The heating system, in turn, is then forced to work longer (or more often) to replace the heat removed by the fan. Also, in addition to removing products of combustion, the draft created in the chimney pulls heated air up and out of the house when the furnace is on—and continues to do so even after the furnace shuts down. Cool air from outdoors replaces the heated air. When the house cools down sufficiently, the thermostat will again signal the furnace to replenish the heat that has been lost.

During a warm summer day, the cycle essentially changes direction. Let's assume the hypothetical house is air conditioned. When the house has warmed up because heat has flowed into it, and because lamps, sunlight, appliances, and people have warmed the air inside, air conditioning forces the unwanted heat out of the house again.

Because you cannot keep heat in the house forever, you must keep acquiring more natural gas, electricity, or fuel oil for the house to use. That, of course, is what costs you money.

When you buy electricity from a local utility company, you pay by the kilowatt-hour. When you buy natural gas, fuel oil, or propane, you pay according to the number of cubic feet or gallons. Most important is the energy available in those fuels. The units that measure energy, and the ones used repeatedly in this book, are the British thermal unit and the kilowatt-hour.

British thermal unit (Btu) is a measure of heat. Specifically, it is the amount of heat required to raise the temperature of one pound of water 1 degree Fahrenheit. (Heat, remember, is a form of energy.) One Btu may seem to be a large amount of energy, but it's not. One gallon of fuel oil would yield about 140,000 Btu if it were burned completely, and 100 cubic feet of natural gas would yield about 100,000 Btu or 1 therm.

Kilowatt-hour (kwh) is a measure of electrical energy. Specifically, it is the amount of electricity required to operate a 1,000-watt appliance for 1 hour, or a 1-watt appliance for 1,000 hours, or any combination of watts and hours that, multiplied together, totals 1,000. A 100-watt light bulb burned for 10 hours would consume 1 kwh.

While electricity is most often used to provide light or motion, it is also used to supply heat. Just as it's possible to measure the number of Btu in a gallon of fuel oil or a cubic foot of natural gas, it's also possible to measure the number of Btu in 1 kwh of electricity. One kwh is the equivalent of 3,413 Btu.

Here and throughout the book, the word *furnace* refers to all types of heating systems, whether oil-fired or gas-fired, including hot-air, hot-water, or steam.

Electric heating systems convert virtually all the electricity they use into heat for the house, but there's a different kind of energy loss associated with electric heat. For every three units of energy that enter a generating plant, only one reaches a customer's home as electricity; the other two units are dissipated when the electricity is generated or along the transmission lines. The generally higher cost of electricity compared to the cost of gas or oil reflects this manufacturing and distribution inefficiency.

IMPORTANT INFLUENCES ON THE "ENERGY SYSTEM"

Humidity, the number of windows—even the location and color of the house—all play a part in determining the amount of energy your house uses. But the three factors that have the greatest effect on the household's "energy system" are temperature, the condition of the house, and the condition of the heating and cooling systems.

1. Temperature—specifically, the difference between outdoor and indoor temperatures. Heat, as we have noted, will flow through solid materials. Heat can move in any direction, and it will always move to an area of lower temperature. The greater the *difference* in temperature, and the less resistance a material offers to the flow of heat, the faster heat will move.

R-value is the term used to express a material's ability to retard the flow of heat. Every part of a house has its own R-value: Window glass has a low R-value; walls have higher R-values depending on, among other things, whether or not they are insulated. The insulation itself has an R-value. Doubling the R-value (by adding insulation to a wall, for example) cuts the flow of heat in half. Some building codes use the term *U-value* to prescribe how effective an insulated wall, ceiling, or floor must be in retarding the flow of heat. The U-value is a measure of heat flow and is the reciprocal of a material's R-value—that is, $R = 1/U$ and $U = 1/R$.

Temperature is measured in degrees; *degree-day* is a term that combines both temperature and time and serves as an index of the energy required to heat or cool a house. During the winter, the greater the number of heating degree-days in your area, the more heat your house will require to maintain a given temperature. Past records of heating degree-days for your area won't tell you precisely how many Btu your furnace will have to produce, but they

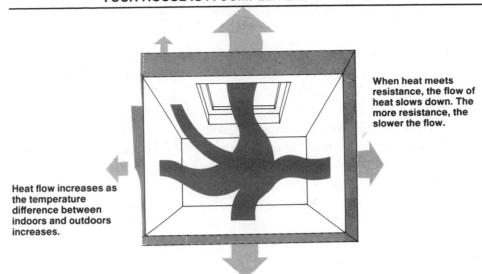

When heat meets resistance, the flow of heat slows down. The more resistance, the slower the flow.

Heat flow increases as the temperature difference between indoors and outdoors increases.

Figure 1.2   Heat flow

can give you a measure by which you can predict the amount of heat you will need. In general, if the number of degree-days drops from one winter to the next, the amount of energy you use for heating should be reduced. During the summer, the greater the number of cooling degree-days, the more air conditioning would be required to maintain a given temperature in a house. (Note, however, that cooling degree-days are a far less useful index than heating degree-days, for reasons explained in chapter 7.)

Heating degree-days (often included in weather reports) are calculated by taking the high and low temperatures for the day, adding them, and dividing the sum by 2. That number is subtracted from 65. For cooling degree-days, the day's high and low temperatures are added and the result is divided by 2; 65 is then subtracted from that number. Why 65? It has been shown that a house generally needs neither heating nor cooling to maintain a comfortable indoor temperature when the temperature outside is 65°F. As a result, 65° becomes a useful starting point for determining the demand that will be placed on the heating and cooling systems.

2. The condition of the house. The house may be what builders call "loose"—that is, there may be cracks around the door or window frames, or around the foundation, and the windows and doors themselves may also be loose or warped. All of these factors contribute to heat losses from air leakage. Conversely, a "tight" house—one that's relatively free of leakage—will permit heated air to escape relatively slowly. But, as indicated previously, even a tight house can lose heat rapidly (by conduction) through the roof, walls, windows, and floors—that is, unless the house has adequate insulation, interior window insulators, or storm windows.

3. The condition of the heating and cooling systems. The inside of the furnace may be covered with soot, which can reduce the amount of heat delivered to the rooms. The oil-burner nozzle or gas orifice may be clogged, air inlets may be improperly adjusted, or the chimney may be obstructed. The furnace controls may need adjustment so that the furnace will fire when it should and shut off when it should. Filters on the air conditioner or furnace may need cleaning or changing. Whatever the problem, it means you are getting less heating or cooling than you are paying for.

At the most elementary level, saving energy in your home means using gas, oil, and electricity sensibly to minimize waste: turning off the lights when you leave a room, for example, or keeping the doors and windows closed when the furnace or air conditioner is running.

To achieve a significant saving of energy, however, keep in mind three broad goals, each of which incorporates the sensible use of fuel and electricity:

■ Getting the maximum amount of energy out of the fuels you buy. Your furnace should be set up and maintained to deliver as much heat as possible from the fuel it burns. That, in turn, will reduce the total amount of fuel the furnace burns over the course of a heating season.

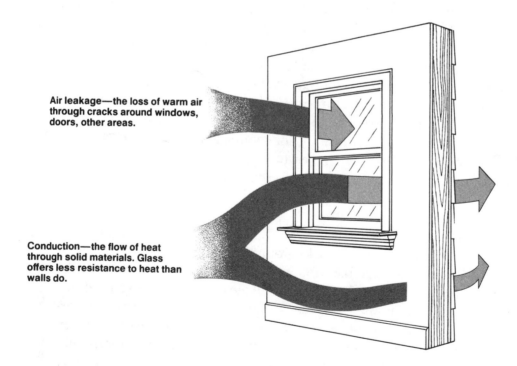

Air leakage—the loss of warm air through cracks around windows, doors, other areas.

Conduction—the flow of heat through solid materials. Glass offers less resistance to heat than walls do.

Figure 1.3 Two ways to lose heat

■ Keeping heat locked inside the house—or, in the summer, locked out—for the longest possible time so that less energy will be required to keep your house at the desired temperature.

■ Tailoring the equipment to the task. In other words, you should choose and use appliances sensibly so that they give you the results you want without using too much energy. In selecting a room air conditioner, for example, look for models with a high energy efficiency rating (EER): The higher the EER, the more Btu of cooling you will receive per watt of electricity consumed. Using an appliance wisely can also save energy. Wash your laundry in warm or even cold water, instead of hot. And always use a cold rinse.

SAVING ENERGY:
A GENERAL PERSPECTIVE

There is more than one way to arrive at these three broad energy-saving goals. To find out which conservation measures will work best for your house, take a close look at your household energy system. Begin with the structure of the house itself to make sure that windows and doors are tight and there are no drafts. Determine whether there is adequate insulation in the walls, ceilings, floors, and crawl space. Consider whether installation of storm windows and weatherstripping could help reduce air leakage—and your heating bills.

Next, look to the heating and cooling equipment and the way you use it. The furnace may be supplying heat when it's not needed—late at night or when no one is home. Changing the thermostat setting, or replacing the thermostat with one that provides automatic setback, may be a useful way to reduce the amount of heating (and cooling, if you have central air conditioning) you use. Of course, if the furnace is operating at less than peak efficiency, you are spending more than necessary to heat your house. A tune-up (by a qualified technician) might save you money. Or you may be able to install certain devices that can extract a bit more heat from the system or keep heat inside the house for longer periods.

Then look to your water heater: There are simple steps you can take to reduce the expense of heating the water you and your family use.

Finally, check your major appliances. Are you using your present equipment to best advantage? When you buy new appliances, are you selecting energy-efficient models?

In attending to these areas—the house itself, the heating and cooling systems, the hot-water equipment, and the appliances—you will be covering almost all the energy consumed in your home. You may well find that there are measures you can take to save energy in all these areas.

Which steps should you take first? No single set of conservation measures will work best for every house. And no single step will *always* result in a significant saving. Obviously, you will want to give special consideration to those measures that will yield the greatest saving per dollar of cost. Unfortunately, generalizations are apt to be of little help to specific homeowners, and the potential saving must be calculated on an individual basis.

The point to remember is this: Because energy consumption varies widely from one house to the next, and because electricity and fuel prices vary widely from one locality to another, the amount of energy and money to be saved will also vary. Investing in insulation may well cut your heating and cooling requirements by 30 percent, but the saving could also be far less than that. You will have to work it out for yourself, and in the chapters that follow, you'll find the information needed to do it.

One approach is to weigh the saving from a conservation measure against its cost, in order to determine the payback period—that is, how long it will take a particular energy-saving step (insulation, storm windows, or whatever) to pay for itself through lower utility bills. Generally speaking, the shorter the payback period, the more attractive a specific conservation measure becomes in relation to others. Let's say a contractor has offered to blow insulation into your walls for $2,000. By using the Insulation Worksheet in chapter 3, you can estimate how much money that insulation will save each year. If the saving will be $400, then the insulation has a five-year payback. But if your calculations show that the same amount of insulation would save only about $100, then you are facing a 20-year payback. At that point, you might want to look for ways to save $100 a year (or more, if possible) for less than a $2,000 initial cost.

TWO

Home Energy Audits

While you can assess your home energy needs on your own with a do-it-yourself audit, your utility company may provide you with a mail-in, computer-processed questionnaire—or even send a trained inspector to your house to offer specific recommendations. The fee, if any, will be nominal.

MAIL-INS VERSUS WALK-THROUGHS

If your utility company offers a home energy audit, find out how much it costs, whether or not it will include an in-home inspection, and, if so, whether it includes testing the efficiency of your furnace. Ask whether the audit will take into account general energy conservation or whether it will be limited to conservation of the fuels that the utility supplies. If the utility does not conduct audits, you may want to call your state energy office (see Appendix A) to find out what options are available to you.

A typical mail-in questionnaire calls for homeowners to fill in basic data about their houses, including the size of the house and its construction, the type of heating system, the sizes and condition of windows and doors, and the amount of insulation in walls, floors, and ceilings. Such forms may also include questions about solar energy.

Once the form has been completed and processed, the homeowner receives a computer printout listing possible ways to save energy. For each item on the list, the printout gives the saving in fuel and dollars, a range of installation costs, and the number of years it should take for each measure to pay for itself through lower fuel bills.

A mail-in audit usually provides only a rough guide to saving energy. If you don't (or can't) answer some of the questions, the computer might try to guess the correct answer for you. If you don't know how much insulation you have in the attic, for example, the computer might assume that your house is "typical." Or, without the needed information, the computer printout might omit any recommendations for added insulation. With an in-home inspection, on the other hand, the auditor can take all the relevant measurements.

If you have a choice between a mail-in questionnaire or an in-house audit, opt for the latter, even if it involves a small cost.

A completed energy audit report should include detailed information about what needs to be done. Estimated costs and savings should be given in dollars and cents. Don't, however, expect the figures in an audit report to be as precise as those in a balance sheet. Instead, consider the report a sort of menu—a list that will allow you to choose what you think will be the most desirable or the most cost-effective options.

Carry out any recommendations that you can afford and that appear to offer reasonable payback (i.e., in less than six years). Many improvements can be adjusted to your preference or lifestyle, but any improvements will help lower your energy bills.

THE PROFESSIONAL AUDIT

The following material about audits has been adapted from information provided by the Conservation and Renewable Energy Inquiry and Referral Service, operated by Advanced Science for the U.S. Department of Energy, and published in March 1989.

A *walk-through* is the most common method of conducting a professional energy audit. It involves a room-by-room examination of a residence, as well as a thorough review of past heating and cooling bills. A walk-through takes note of: the amount of insulation in attics, crawl spaces, unconditioned basements, and other visually accessible parts of the building envelope; type of building construction; types of windows and doors; the occupants' energy-use habits; and other factors that may affect energy consumption.

More complex audits often use tools such as blower doors, infrared cameras, and digital surface thermometers to detect leaks. A *building pressurization test* measures the leakiness of the building, and a *thermographic inspection* reveals the often-hard-to-detect areas of infiltration and areas where insulation is missing. However, these tests tend to be expensive, so most homeowners do not invest in such detailed audits.

What to Do When Your Home Is Being Audited

1. Make a list of any problems, questions, or improvements that you might want to discuss with the auditor. Assemble copies or a summary of the year's energy bills for all fuels. Make notes of occupant behavior. Some questions to consider are: Is anyone home during working hours? What is the average thermostat setting during the summer? During the winter?

2. The auditor most likely will measure spaces; check the condition of caulking, weatherstripping, and insulation; and look for areas of infiltration. As you go along, ask any questions you may have. The auditor may use a smoke pencil to demonstrate air leaks.

3. Ask the auditor what the findings indicate. Does the auditor have any suggestions for improving the energy efficiency of your home, such as recaulking, adding storm windows, or putting in additional insulation? What would be the cost and payback periods of any improvements suggested? Does the company offer any financing or other incentives for the more expensive retrofits?

Thermographic Inspection

Thermography measures surface temperatures using specifically designed infrared cameras with heat-sensitive film. Images on the film record the temperature variations of the building's skin, ranging from deep red for warm regions to blue for cold regions. Thermograms of electrical systems can detect abnormally hot or cold electrical connections, wires, or components. Energy auditors use thermography to detect thermal radiation, conduction, and air leakage from a structure.

Infrared scanning can be used to check the insulation in a building's walls. The resulting thermograms help auditors determine whether or not insulation is needed—or, as quality control, to ensure that insulation has been installed correctly. In addition to the use of thermography during an energy audit, a scan can be performed before the purchase of a house; even new houses can have unexpected problems. Sometimes a purchase contract contains a clause requiring a scan of the house to document how well the house is insulated.

Infrared scanning consists of either an interior or an exterior survey. An exterior survey has a number of drawbacks. Warm air escaping from a building does not always move through the walls in a straight line. Heat loss detected in one area of the outside wall might originate at some other location on the inside of the wall. Air movement also affects the assessment of temperature differences. The greater the wind speed, the harder it is to detect temperature differences on the outside surface of the building. Thus, interior surveys may be more accurate because of the lower level of air movement.

Building Pressurization

Pressure tests use a blower door to determine the susceptibility of a home to air infiltration. Several reasons for establishing the proper tightness of a building are: to reduce energy consumption due to air leakage; to prevent condensation problems; to prevent drafts caused by cold air leaking into the airspace; and to assess the need for mechanical ventilation.

A blower door is a powerful fan that mounts into the frame of a door leading outside. It is important that the blower door have access to the outside so it can draw air from the interior of the house. When the fan is turned on, it pulls air forcefully out of the house, creating a partial vacuum inside the house. The lower inside air pressure will then draw outside air in through all unsealed cracks and openings. Blower doors are either *calibrated* or *uncalibrated*. The calibration device on the fan measures the amount of air flowing through the house and out through the fan. Uncalibrated blower doors can be used to help locate leaks in homes, but a calibrated door provides more specific information that can be used to evaluate the overall airtightness of the home.

THE DO-IT-YOURSELF AUDIT

When auditing your own home, first make a list of obvious leaks. Note any broken windows and be sure to check basement windows for looseness or broken panes. Determine whether fireplace dampers, basement doors, attic doors, and doors to attached garages close tightly. There are many potential sites for air to leak into or out of your home, including: windows and doors, gaps around pipe and wire feedthroughs, electrical outlets, foundation seals, mail slots, exhaust fans, attics, garage doors, siding cracks, and old caulking. Below are some simple steps for conducting your own audit.

Room-by-Room

Check for indoor air leaks, such as gaps along the baseboard or edge of the flooring and at junctures of the walls and ceiling. See if air is flowing out through electrical outlets and switchplates. Note whether exterior walls are insulated. Compare the R-value of your home's insulation with the minimum R-value required for your region.

Inspect windows and doors for air leaks. See if you can rattle them—movement means possible air leaks. If light shines through door and window frames, or if a piece of paper slides easily between the frame and the door or sash, the door or window may be leaky. These leaks can be sealed by caulking and weatherstripping. Note whether existing caulking and weatherstripping has been properly applied and is in good condition.

If you are having difficulty locating leaks, you may want to conduct your own building pressurization test. First, close and lock all doors, windows, and fireplace flues. Then turn on all ventilating fans (usually in the kitchen and bathrooms). This will increase infiltration through cracks and leaks, making them easier to detect. Punk, incense sticks, or your damp hand will work well for locating these leaks. Moving air will cause the smoke to waver, and a breeze will be felt easily as it cools your damp hand.

The attic. If the attic hatch is located above a heated or cooled area, see if it is insulated and weatherstripped. Check whether openings for items such as pipes, ductwork, and chimneys seem to be sealed properly.

During any subsequent work, gaps should be stuffed with insulation and caulked whenever possible. However, never cover light fixtures with insulation. To avoid overheating, allow a 3-inch space around these fixtures.

Check to see if there is a vapor retarder under the attic insulation. If there isn't, you may want to consider painting the interior ceilings with vapor-retarder paint to reduce the amount of humidity that passes through the ceiling. Moisture can reduce the effectiveness of insulation. If there are attic vents, be sure they are not blocked by insulation.

The basement. If unheated basement areas are located under living spaces, determine whether there is insulation under the flooring. Inspect the caulking and insulation between the basement and the floor of living spaces. Where the top of the foundation meets the floor, insulation should have an R-value of 11 or greater. If the basement is heated, the exterior walls should be insulated. Check windows for broken panes, cracks, or loose frames. Note whether all hot-water pipes and ductwork are insulated. See if caulk or insulation is needed where the house framing meets the foundation. Look for water seepage on basement walls; if no seepage problem exists, walls in heated basement areas can be insulated. Look for insulation between heating ducts and outside walls, especially where ducts bend to feed first-floor registers. Inspect for insulation around flues or plumbing vents that connect to the attic.

Outside

Inspect the caulking wherever two building materials meet. Look for cracks in places such as the mortar, foundation, and siding. Other places to look for leaks include: all exterior corners; the spaces between chimneys and siding or brick; areas where the foundation meets the bottom of exterior brick or siding; and holes or service entries for faucets, pipes, electrical outlets, and wiring. Check exterior caulking around doors and windows, and see whether exterior doors seal tightly. See if air is escaping around or through your wall or window-mounted air conditioner(s). During the winter, remove window-mounted air conditioners or use plastic covers or clear plastic sheeting to seal air conditioners against air leakage.

Inspect for broken glass or loose-fitting storm windows. See if triple-track windows need lubrication or adjustment.

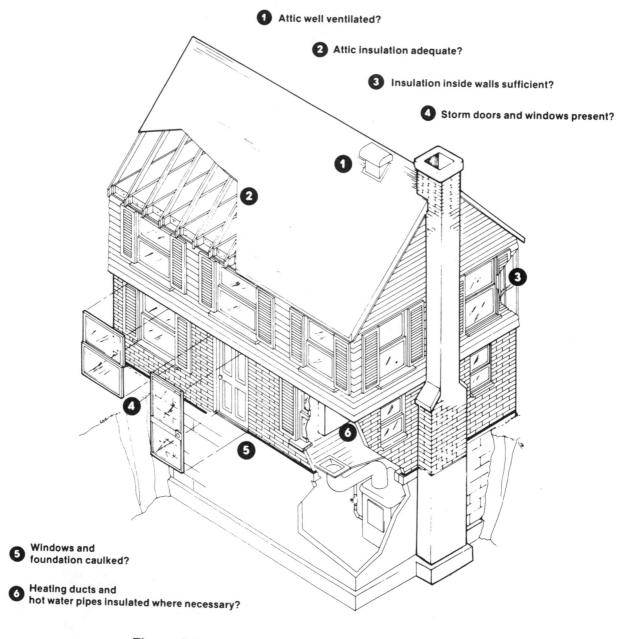

1 Attic well ventilated?

2 Attic insulation adequate?

3 Insulation inside walls sufficient?

4 Storm doors and windows present?

5 Windows and foundation caulked?

6 Heating ducts and hot water pipes insulated where necessary?

Figure 2.1   Six highlights of a home energy audit

THREE

Keeping Heat Leakage
to a Minimum

House heat escapes through every leak. You could keep the house warm by adding more heat, but heat is expensive, so it's better to plug the leaks by weatherizing the house.

Cracks and holes in a house's "thermal envelope" (the exterior walls, roof, and floor) can develop at any time and in any house—new or old. They can be caused by poor construction, settling and aging of the house, and wide variations in temperature and humidity. They are especially likely to develop where two different surfaces meet, such as the wall–foundation or chimney–wall joint. As the weather gets colder, air passes in and out of these cracks and holes. A large discrepancy between indoor and outdoor temperatures encourages warm air to migrate. Winter winds force dense, cold, outdoor air through the openings and into the house.

According to the U.S. Department of Energy, air leaks account for one-third of the heat lost from a house that has not been weatherized. Thus, making a house as leakproof as possible is a reasonable first step toward saving energy and reducing fuel costs. Fortunately, it is also one of the least expensive energy-conservation measures, although it does require some effort on your part.

Another kind of leak that affects home heating bills results from conductive losses—the flow of heat through solid materials. Reducing conductive heat losses is more complicated than sealing air leaks. Windows lose a good deal of heat by conduction, as do walls, doors, the roof, and any other surface that is cold on one side and warm on the other.

You can take a number of effective measures to reduce air leakage. All the methods in this chapter can save you energy and money, but there are also a number of side benefits. Tightening a house (with thorough weatherstripping, caulking, and/or storm windows) will make it more comfortable: A tighter house will seal in some of the moisture from bathing, washing, and cooking, which will help to provide a more comfortable humidity level during the heating season. Drafts, too, will be reduced.

Some methods offer special extras: Weatherstripping, for instance, acts as sound insulation, reducing the noises of doors closing and windows rattling. And it helps keep out dust and soot. Caulking tends to preserve the structure of your home by keeping water out of cracks and joints. Insulation can help keep walls and floors warm. Although these benefits may not be the main reason you would weatherstrip, caulk, or insulate your house, they provide an additional incentive.

WEATHERSTRIPPING

Money in the form of heated air escapes through loose windows and doors. Consumers Union found that a fairly loose double-hung window lost $6 a year. Weatherstripping the window reduced the air infiltration by half.

A dozen or so types of weatherstripping can be bought in lumberyards as well as in hardware, discount, paint, department, and variety stores. Many of the products come in long rolls designed to be cut into lengths to fit into place. With few exceptions, all can be used on doors as well as windows. All allow doors and windows to operate normally. The products can be applied from the inside or the outside: Some can be tacked on and some are stuck on with their own adhesive. Some are nearly invisible when in place; others are all too visible.

Double-hung windows should have weatherstripping placed around the entire frame and also where the two sashes meet. Casement windows and doors should have weatherstripping all around the frame. (Bottoms of doors present a special problem—and need special treatment. See "Special Weatherstripping," later in this chapter.)

Varieties of Weatherstripping

Here is a brief summary of the various types of weatherstripping:

Nonreinforced, self-adhesive foam. Probably the most common type of weatherstripping, these floppy, nonreinforced strips usually are inexpensive. Some are polyurethane ("open-cell") foams, soft and lightweight. Some are vinyl ("closed-cell"), made of a somewhat denser, firmer material. Some are black sponge rubber, which is even denser than vinyl and available in a greater selection of thicknesses. All of these foams have a bubbly texture and a smooth surface. Another foam, made of rubber with a firm, nonporous surface (sometimes ribbed) is more expensive than the other types.

Self-adhesive foam is possibly the easiest weatherstripping to install. It should be applied to a smooth, dry, dust-free surface that isn't too cold. You must be careful not to stretch the foam while you are putting it on, as it will pop off when it begins to shrink back to size.

All the nonreinforced foams, particularly the thicker ones, resist being straightened out from their rolled-up position. As a result, they are best used only on the insides of door jambs or on window tops and sills, where they usually are held in place by compression. Foam strips are invisible when they are installed at the tops and bottoms of window sashes. On a door jamb, they are visible only when the door is open.

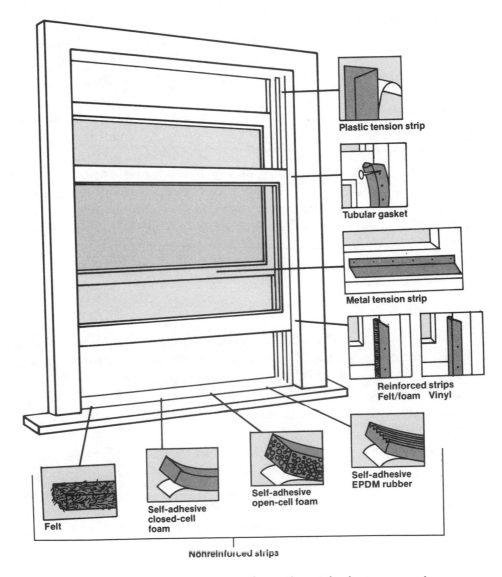

Plastic tension strip

Tubular gasket

Metal tension strip

Reinforced strips
Felt/foam   Vinyl

Self-adhesive
EPDM rubber

Self-adhesive
open-cell foam

Self-adhesive
closed-cell
foam

Felt

Nonreinforced strips

Figure 3.1   How types of weatherstripping are used

If these foams are placed between sliding parts, the friction may abrade them. For example, they should not be placed where the two sashes of double-hung windows meet.

When you have to replace foam strips, you may find them difficult to remove. After two years or so, the adhesive may have become crumbly and the foam may pull off in chunks rather than in a nice, neat strip.

Nonreinforced felt. Inexpensive, nonreinforced felt strips are the "traditional" type of weatherstripping. They come in polyester or wool in a variety of widths and are not self-adhesive. The strips have to be nailed, stapled, or glued in place. Felt can be used in much the same manner as foam; unlike foam, however, it can resist abrasion. Because felt tends to hold moisture, it's a good idea to use it only where it is not exposed to the weather.

Reinforced vinyl, felt, and foam. These strips usually are reinforced with a fairly stiff strip of aluminum or plastic. Some of the foam types come in straight lengths rather than rolls, and they have a wood backing. Reinforced stripping is butted against closed windows or doors, inside or out, and then stapled, nailed, or glued to the frame. It can be used with all types of doors and windows. If the window is one you want to be able to open, reinforced vinyl or felt makes an especially useful seal because it can be fastened firmly to the window frame; reinforced foam is far less durable for this use.

Reinforced weatherstripping usually is quite visible when it is in place. Some brands come in colors, and some are even wood-grained, which helps them blend with their surroundings.

Tubular gaskets. Some tubular gaskets are hollow and flexible; others are foam-filled and rather firm. They are often made of vinyl or rubber. Like the other reinforced strips, gaskets are fully visible when installed. The tubular section of the gasket butts against the closed door or window. The flat lip, or flange, is nailed or stapled to the frame.

Tension strips. These thin, flexible products bend or fold in place to form a tension seal. Most are metal, generally packaged with nails for installation. Others are made of self-adhesive plastic. If invisibility is important, tension strips usually are your best bet. These strips fit into the side channels of a window and at the tops and bottoms of both sashes, where they are completely out of sight when the window is closed. Tension strips are equally unobtrusive when they are fitted into the jambs around a door. They are the only weatherstripping that can successfully close the gap where the sashes of double-hung windows meet.

With metal strips, one side is affixed to the frame; the other is bent out just enough to form the tension seal between the door or window and its frame.

The plastic V-strip adheres along one edge; the other edge is folded over to make the V. The seal is formed by the pressure of the V trying to spring open.

Selection

If properly installed, just about any type of weatherstripping should stem air leakage well. But that's not to say that the various types are interchangeable. It will pay you to pick the right type for your installation. Here are some factors to consider in making your selection.

Resistance to weathering. Among foam weatherstrips, vinyl is likely to be most durable. It is the least likely to discolor, lose flexibility, or come unglued. Rubber foams also are quite long-lasting, but they may harden and lose flexibility. Sponge-rubber foams hold up almost as well. Polyurethane foams tend to shrivel, discolor, and crumble in time.

Compression. Foams, felts, and gaskets seal by compression. A material that compresses well can fill in gaps in irregular surfaces and thus form a tight seal.

Thin tension strips, of course, cannot fill irregular gaps. Among the various reinforced and nonreinforced foams, polyurethanes are the most compressible; the vinyls run a close second. Rubber strips are the least compressible of the foams, but they are more compressible than felt stripping.

Water absorption. Felt is the most absorptive. Polyester dries readily and wool tends to stay damp. Such moisture could easily foster mildew or rot.

Polyurethane foam absorbs a lot of water. Vinyl foam and sponge rubber absorb less. And foams dry quickly. Rubber strips won't absorb any noticeable amount of water.

Special Weatherstripping

For specific problem areas, special-purpose weatherstripping materials are available.

Door sweeps, door bottoms, and thresholds. Two types of weatherstripping have been designed to deal with the gaps at the bottoms of doors—often a major source of heat loss. Door sweeps are attached to the inside face of an in-swinging door (or to the outside face of an out-swinging door) with just enough overlap at the bottom to seal the gap. Most sweeps contain a felt, vinyl, or brush strip reinforced with aluminum or wood. They are easy to install; most just screw on. Ordinary door sweeps work well unless there is interference because the door has to swing open over exceptionally deep carpeting. If so, automatic sweeps are available to solve this problem: They rise automatically when the door is opened and lower when it is shut.

Door bottoms attach to the bottom of a door to fill the gap between door and threshold. Most are vinyl gaskets held in place by an aluminum or wood channel that screws onto the door. There are two types, one far easier to install than the other. The easy version is an L- or U-shaped channel that fits over the bottom of the door but can be screwed into the face of the door. This type generally has slotted screw holes that allow you to adjust the height after the device is installed. The channel holds a vinyl or other pliable gasket that meets the sill when the door is closed. This type is somewhat visible when in place. More difficult to install but less visible is a gasket that requires removing the door—it attaches to the bottom edge.

Threshold weatherstripping is attached to the sill beneath the door, replacing the existing threshold. Typically, this variety is made of aluminum or wood, with a channel containing a rubber or vinyl gasket. The only drawback to the threshold type of weatherstripping is that the gasket can be worn down quickly if the doorway is in constant use. While a gasket can be replaced, so you don't have to buy a whole new threshold, gaskets may be difficult to find in local stores.

Gaskets for wall outlets. Electrical wall outlets and wall switches can be an unexpected avenue of air leakage to and from the outdoors. Even those mounted on interior walls often are connected to an unheated area via a hollow wall that lets drafts circulate. These leaks account for a surprisingly high percentage of a house's total air leakage. You can buy special gaskets that quickly and inexpensively block the air that sneaks around electrical outlets and light switches.

| Threshold | Door sweep | Door bottom |

Figure 3.2   Special weatherstripping

There are several brands of gaskets available; you'll probably find at least one brand at your local hardware store. To be on the safe side when installing a gasket, first remove the fuse or open the circuit breaker on the appropriate circuit.

Recommendations

Weatherstripping won't compensate for major gaps where a window sash fits into its frame. Such a window needs to be fixed or replaced. Nor does weatherstripping eliminate the need for storm windows. Nevertheless, selecting the appropriate weatherstripping will prevent or reduce the loss of heat due to air leakage in your home.

A rule of thumb in buying weatherstripping is to decide on the type that's best for your use, regardless of brand, and then price-shop. There won't be much quality variation among brands of a single type.

If you want weatherstripping that is invisible once it's in place, look to tension strips or, for certain applications, self-adhesive foams. But expect adhesion problems with some foams, especially when temperature fluctuates seasonally. It's best to use the foams where they are held in place by compression and where they aren't subject to friction or abrasion.

Closed-cell vinyl foams and tension strips are the longest-lasting type of weatherstripping, but they tend to be expensive. They can be used for just about every kind of door and window joint, including that difficult spot where the two sashes of a double-hung window meet. The plastic self-adhesive tension V-strips are easier to install than the metal ones, but both types of weatherstripping should last a long time. The self-adhesive, prefolded tension strips are particularly convenient to install around doors.

Putting in reinforced felt or foam strips and tubular gaskets requires about as much work as nailing in tension strips. Most types of reinforced weatherstripping will last several years. Probably the biggest drawback with felt or foam strips and tubular gaskets is aesthetic: They are always in full view.

Nonreinforced felt can be installed fairly easily with a staple gun or a tack hammer. (If the stripping is to be exposed to moisture, however, use non-rusting staples or tacks; ordinary fasteners rust quickly.) Because nonreinforced felt is cheap, it may be the right choice if you have to seal great lengths where appearance is secondary.

If you have windows or doors that will be kept closed all winter, there is still another choice: tape. Both transparent plastic and duct tape will hold up well. They aren't pretty, but they are inexpensive. With attic or garage windows, for example, tape might be perfectly adequate, particularly if it is installed on the inside. One drawback to consider: When the tape is removed, it might pull up loose paint or leave a residue of adhesive that is hard to remove.

EXTERIOR CAULKS

In the past, houses were caulked mainly to protect the structure from water seepage and to reduce drafts. Householders now realize that caulking is also a key to saving fuel dollars. Like weatherstripping, caulking helps save energy by reducing cold-air leakage. That cold air not only is an extra burden for your heating system but also can make a home uncomfortably drafty.

Caulking's triple payoff in waterproofing, energy savings, and comfort may be gratifying, but the job itself is no fun. You have to strip out crumbled old caulk, gun a bead of new compound evenly into cracks and joints, and perhaps tool the caulking down flush with the siding. You may also want to paint over the caulk so it matches the color of your house. Because caulking is not a chore you'll want to take on more often than necessary, choose a caulk that applies easily and adheres well.

While many caulks are formulated for the construction industry, only a few types are marketed to do-it-yourselfers: latex-based, silicone-based, and latex/silicone hybrids. (Oil-based, butyl, and urethane-based caulks are not in demand for the home market.)

Home caulks cure to a rubbery solid that expands and contracts with changes in temperature. This elasticity helps them maintain their seal through seasonal changes.

Latex and acrylic caulks generally hold a coat of paint well. Their directions usually recommend letting the latex products set for at least a half hour before painting, but you probably can get away with painting them immediately after application if you are careful not to deform the caulk. Being able to caulk and paint in almost a single operation is a real plus if you're working on a ladder.

Old-style silicone-based caulks normally were not paintable—paint just beaded up on them—but recent formulations have been produced to deal with that problem. Some "paintable" silicone caulks now can hold a coat of paint, though not necessarily very well. Manufacturers are faced with the concern that adapting a formula to take paint may compromise the product's weathering properties.

Where to caulk. The obvious places to caulk are around door and window frames and at the corners of the house. But other important places also may need to be sealed. Check first for gaps between the house framing and the foundation. To spot these gaps, stand inside the basement and look along the top of the basement wall. If you can see light, you need caulk. Another place to check is the space between the siding and the sheathing. Caulking behind the siding helps block the flow of air so the house can retain its heat longer. As we've noted, cold air can find its way into the house even through such unlikely inlets as electric switches and receptacles.

How to caulk. Amost every type of exterior caulk comes in a cylindrical cartridge about a foot long. Most people use a simple hand-operated caulking gun to squeeze a bead of the stuff out of the cartridge. The instructions for some brands suggest that bare wood surfaces should be primed before caulk is applied. As a general rule, priming any surface before caulking is likely to help; if you are applying caulk to a metal surface, it may help a lot. Here are some tips for working with caulk:

■ A clean surface is essential. Remove old caulk. Be sure that joints are dry (only latex caulks can be used in damp joints). Latex can be cleaned up with water; with other caulks, you can aid cleanup by pressing masking tape next to the crack before caulking. Just pull up the tape after caulking and the excess caulk comes up with it. Don't caulk if it's colder than 45°F.

■ Cut the tip of the cartridge nozzle at an angle so you can slant the cartridge for better visibility and control. The nozzle cut should just span the joint you are trying to fill (⅛- to ¼-inch diameter usually is enough). You can fill extra-large gaps by moving the tip from side to side.

■ Poke the nozzle into any joints wider than the nozzle so that the caulking fills the gap from the bottom up. Rough-fill the bottom of a deep, wide gap with oakum, cotton, or glass wool, or with another inert filler, to within ½ inch of the surface.

■ Press the nozzle firmly over all joints and move it so that the caulk is pushed ahead of the tip. That way, you'll force the caulk deeper into the crevice and reduce the nozzle's tendency to pull the caulk up and out of the joint.

■ Before you level and trim the caulking, close off the cartridge. Push a large-headed nail (a common tenpenny nail serves well) into the nozzle and wipe away the excess that oozes out. Cover the nail and nozzle end with aluminum foil. Finish the cartridge within one season to prevent the material from drying out and becoming useless.

■ Remove excess caulk and smooth the joint with a moistened stick, putty knife, or rubber-gloved finger along the surface of the joint. Immediately remove excess caulk at the sides with a cloth you wet with water (or recommended solvent). Remove any masking tape before the caulk has had a chance to set.

Safety

The labels on a number of products recommend avoiding eye and skin contact with uncured caulk. Those on a few brands warn of solvents or other agents that may be harmful to inhale. If the product you use emits such harmful vapors, adequate ventilation is a must when caulking indoors. With any caulk, you should wear gloves and wash up thoroughly afterward.

Recommendations

Conventional wisdom has held that silicone caulks are the best—costly, perhaps, but more durable than other caulks. Consumers Union tests have questioned that conclusion. Some silicones hold up no better than many cheaper latex formulations. If you are planning to paint the caulk, the choice is clear. A latex-based caulk usually is more satisfactory when painted than a "paintable" silicone.

Don't buy a "clear" caulk and expect an invisible bead. It may be colorless once it cures, but it won't be more transparent than, say, petroleum jelly. There are also colored, or pigmented, caulks that are tinted to match popular colors of siding. The pigmented version of a given brand and type is apt to adhere and weather as well as its white or clear counterpart, and it might be more unobtrusive when in place.

Whatever your choice of caulk, don't put too much stock in warranties. You may find 10- or 20-year warranties, or even one for the "lifetime" of your house while you own it. But such warranties usually are so bound by conditions that the price of getting them honored is likely to be greater than the satisfaction you might exact, which never amounts to more than a bare refund or a replacement cartridge of the very caulk that left you dissatisfied in the first place.

STORM WINDOWS

Loose windows cost you money. Your furnace must run longer than it would otherwise need to in order to replace the heat that escapes through the windows. Caulking and weatherstripping the *prime windows* (the ones installed when a house is built) will start you on the road to saving energy and money. Installing storm windows will increase the total saved.

The amount of energy and money you could save with storm windows will depend on several factors: the number of windows in your house, and their size; whether you heat your house with oil, gas, or electricity; and the annual number of degree-days in your area. You can estimate your saving with the help of the worksheet at the end of this section on storm windows.

Although just about any storm (or thermal) window can help you save energy and money, the way a storm window is built and installed can make a difference in the amount you can save. Of course, not all prime windows are created equal, either. And even a top-quality window will suffer over the years: The frame and sash can warp or shrink; caulking and weatherstripping can deteriorate. The result is air leakage around the sash and frame.

A storm window, installed over a prime window, serves two purposes: (1) The air space between the prime window and the storm window provides a measure of insulation against the loss of heat by conduction (the transfer of heat through the glass itself). A storm window can cut conduction losses by about half. (2) A well-constructed, tightly installed storm window reduces the amount of air that leaks past a prime window that's not in very tight condition.

The storm window thus becomes half of a window "system." To save the greatest possible amount of energy, both the prime and the storm windows should be tight. That way, the prime window and the storm window together form a rigid, well-sealed barrier against the wind and the cold.

That principle has always been an accepted bit of energy conservation wisdom. Until about 20 years ago, however, little research had been done to determine the range of savings possible with various storm window/prime window combinations. Research in the 1970s, including Consumers Union's tests for a 1980 report, indicated that some widely used data on window performance tended to overstate the amount of energy a storm window could save.

Rating storm windows by brand name is impractical because the storm-window trade is virtually a cottage industry. Several firms build and sell components, and final assembly is done by small companies. As a result, while it is impossible to single out specific brands as good or bad, it is possible to offer practical advice on finding high-quality storm windows.

Triple-track models are the most popular variety. A triple-track storm is permanently installed over the outside of an existing prime window, thus eliminating the bother of putting up screens in the summer and replacing them with storm windows in the fall. A triple-track window has two panes of glass, each of which slides up and down in its own channel, or track, just as the sashes do in a double-hung prime window. The third track holds a screen for the lower half of the window. During the winter, the screen section is

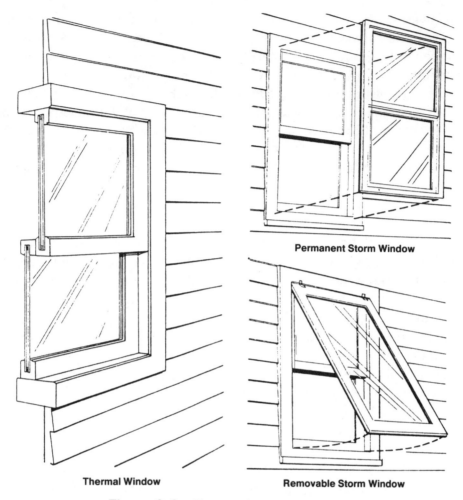

Thermal Window

Permanent Storm Window

Removable Storm Window

Figure 3.3 Types of storm windows

raised and one glass sash is lowered, to form the full storm window. In the summer, for ventilation, one sash goes up and the screen comes down.

Other styles include double-track models with a steel frame, or a triple-track type with a frame of plastic rather than aluminum.

Window Leaks

In Consumers Union tests, the most expensive storm window was also the tightest. It had a leakage rate of 1.2 cubic feet of air per minute (cfm) for each foot of *crack*—the gaps between sash and frame and where the two sashes meet. And one of the cheapest windows was the loosest, with a leakage rate of about 4.35 cfm per foot. Don't assume from these figures, however, that more money always buys a better storm window. Except for the best and the worst, there was no correlation between price and tightness.

The leakage rates are cited here primarily to show that substantial differences exist among storm windows. But knowing how well a storm window controls air leakage will not, of itself, tell you how well the storm window and the prime window together will perform.

If the prime window is the tighter of the two, it will do more to reduce air leakage. If the storm window is the tighter window of the pair, then the storm window will reduce air leakage more. But even the worst storm window, tested by Consumers Union in conjunction with a "loose" prime, cut leakage for the window system by about 7 percent.

Conduction and air leakage must be considered when assessing how effectively storm windows reduce total heat loss. Most storm windows reduce the rate of total loss by about half. (That's to be expected: The conductive losses of a window system are much greater than the air-leakage losses.)

Mist on a storm window is a sign that the window is doing its job. It's also a sign that the prime window is loose, thus allowing warm, moist air to escape from the house. To prevent the mist, you should putty and weatherstrip the prime window.

Shopping for Storm Windows

It pays to shop carefully for storm windows, rather than take what a dealer has in stock or buy on the basis of a come-on price. A dealer may stock windows made by two or three different companies. Some dealers may have one company's wares displayed in the store but then try to sell you another company's products. Don't be steered away from low-priced storm windows just because the salesperson says, for example, that the windows are for an apartment and that you would not want to put them on your house. Inspect the windows closely. Figure 3.4 shows a typical storm window and its components. Here are the important details to look for:

Rigid frame. The pieces of the frame should be put together securely, so that the sides don't blow away from the sashes and the frame doesn't twist. With a rigid frame, a storm window will be better able to resist wind pressure. Corner joints should be neat and solid, with no cracks where the pieces come together. On some models, the corners are mitered—each piece is cut at a 45-degree angle, then joined to form a square corner; on other models, the pieces overlap at the corners. In our view, the quality of the construction matters more than the method of construction.

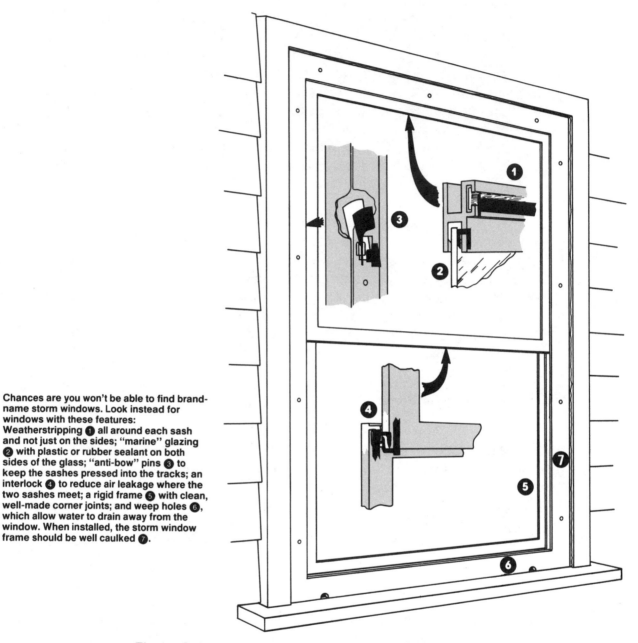

Chances are you won't be able to find brand-name storm windows. Look instead for windows with these features: Weatherstripping ❶ all around each sash and not just on the sides; "marine" glazing ❷ with plastic or rubber sealant on both sides of the glass; "anti-bow" pins ❸ to keep the sashes pressed into the tracks; an interlock ❹ to reduce air leakage where the two sashes meet; a rigid frame ❺ with clean, well-made corner joints; and weep holes ❻, which allow water to drain away from the window. When installed, the storm window frame should be well caulked ❼.

Figure 3.4 What to look for in a storm window

Tie bar. This stiffener connects the sides of the frame at the middle, where the sashes meet. It helps prevent the sides of the window from bowing apart.

Weatherstripping. Typically, only the sides of each sash will have weatherstripping. Sometimes, the stripping is installed in the tracks rather than on the sash itself. It would be better to have weatherstripping all around the sashes to block air leakage at the top of the window and at the sill.

"Marine" glazing. Look for a bead of plastic or rubber sealant all around the edge of the glass on both sides of the pane. This "marine" glazing makes for a stronger, tighter sash. With "drop-in" glazing (the other commonly used method of sash construction), sealant is applied to only one side of the glass.

Tightly fitting sashes. The glass and the screen should fit snugly into their tracks. On a well-constructed window, you should not be able to jiggle a sash up and down when it's locked in position.

Sash interlock. Where the two sashes come together, they should physically lock together—that is, a flange along the top of the lower sash should interlock with a similar flange along the bottom of the upper sash. To test how well this interlock works, close the window and try to force the two sashes apart. If you can separate them, it's a sign that one of the two sashes doesn't sit quite right in its track and thus cannot interlock with the other sash. Some windows have weatherstripping, rather than an interlock, where the two sashes meet. That could be as effective as an interlock, provided the rest of the window is rigid.

"Anti-bow" pins. These are small wedges that help keep the sashes pressed tightly into their track, as a further protection against wind gusts. If the storm windows you want don't have them, you can improvise by wedging small blocks of wood into each track.

Weep holes. The storm-window frame should have two small holes (roughly ¼ inch in diameter) along the sill. They are intended primarily to allow water to drain away from the inside of the storm-window frame during the summer, when the screen may let rain into that area.

Installation

Even the best storm window won't be an effective energy-saver if it's installed badly. Installing triple-track storm windows can be a do-it-yourself project, albeit a time-consuming and physically demanding task. It's important to measure your prime windows carefully so the storm windows fit properly. And while the first window you carry up a ladder might not seem heavy, the tenth or twentieth one almost surely will.

Unless you are reasonably handy with tools—and not daunted by the prospect of spending several hours up on a ladder—you probably will want to hire a professional to install storm windows, or buy them with installation included. Installation costs vary widely, just as prices for storm windows do.

In order to minimize problems that might arise between you and the installer, be sure to take the precautions outlined in chapter 9 for dealing with contractors.

IMPROVING YOUR WINDOWS' ENERGY EFFICIENCY

Improving your windows can be an easy, inexpensive way to save energy. Remember that simply locking the windows during the winter can help keep out some drafts. If you want to achieve more than that, inspect your prime windows and make improvements where necessary. There should be no gaps or cracks in the caulking, and the weatherstripping should be in good repair. If you do nothing else to your windows this year, at least be sure that the caulking and weatherstripping are adequate. The same goes for any storm windows you might already have. Figure 3.5 shows how to replace the weatherstripping on a storm window. Storm windows require less attention than other types of insulators. That's one reason they may well be the product of choice for most homeowners.

When you shop for storm windows, inspect them carefully to be sure they contain the important construction details shown in Figure 3.4. To estimate the benefit that storm windows can mean for you, use the Insulation Worksheet that appears later in this chapter.

If you have bare windows, you may want to consider other forms of insu-

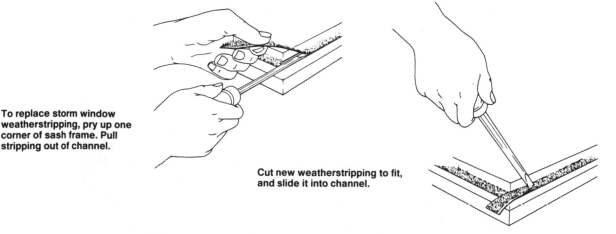

To replace storm window weatherstripping, pry up one corner of sash frame. Pull stripping out of channel.

Cut new weatherstripping to fit, and slide it into channel.

Figure 3.5 How to install weatherstripping in storm windows

lation than storm windows. Among these, insulated shutters and shades cut heat the most, but they require a good deal of time and care. You gain considerable heat by day if you let the winter sun shine in, but to pick up that free heat and retain it, shutters or shades on windows facing east, south, and west should be opened or closed according to the time of day and the level of sunlight. Such careful management takes some effort.

Shades and draperies are useful for windows that are too large to be covered with conventional storm windows, or for skylights and other glassed-in areas easier to insulate from inside.

Remember that the first layer of insulation you add will yield the greatest saving. Piling on even more insulation will not yield a proportionately higher saving. If you increase a window's R-value from 0.9 to 1.8 by adding a storm window, you could reduce heat loss by half. You would have to raise the R-value from 1.8 to 10 to get a comparable dollar saving. For that reason, it may be wise to look for other ways to save energy if you already have storm windows in place. The Insulation Worksheet in this chapter can help you figure out what will yield the greatest saving of energy and money.

For more information about improving the energy efficiency of windows, you may want to obtain the United States Department of Energy Fact Sheet FS216 (Conservation and Renewable Energy Inquiry and Referral Service, P.O. Box 8900, Silver Spring, MD 20907, or call toll-free 1-800-523-2929).

HOW TO CALCULATE SAVINGS TO BE GAINED BY MODIFYING WINDOWS

By using the Window Saving Calculator Worksheet, you can calculate the amount of fuel cost saved by installing storm (or thermal) windows. You can also use part of the worksheet to estimate the saving from weatherstripping alone. Included, as examples, are calculations for installing weatherstripping and permanently mounted tight storm windows and for weatherstripping alone on a hypothetical house in Columbus, Ohio. Here's what to do:

1. Determine the total square footage of the windows in your house, then enter this amount on the Window Saving Calculator Worksheet later in this chapter. To calculate square footage easily, measure the height (in inches) and the width (in inches) of each window. Multiply the two numbers, then divide by 144. Repeat for each window, then total all the areas. (The Columbus house has 280 square feet of window area.) Enter the window area for your house in column 1.

2. Find the conductive heat saving. The accompanying map divides the country into eight heating degree-day zones. Find the zone for your area, then

turn to the table on page 35. The numbers in that table show, for each zone, the heat units (thousands of Btu) saved per square foot by installing various types of windows—storm, double-glazed thermal, or triple-glazed thermal windows.

Move down the left-hand column to find the type that most closely matches the windows you now have. Then read across the table until you come to the column that most closely matches the type you want to install. What you will find is the approximate amount of heat saved per square foot. In the example, it is assumed that the house has single-glazed, double-hung windows, which will be augmented with permanent storm windows. The

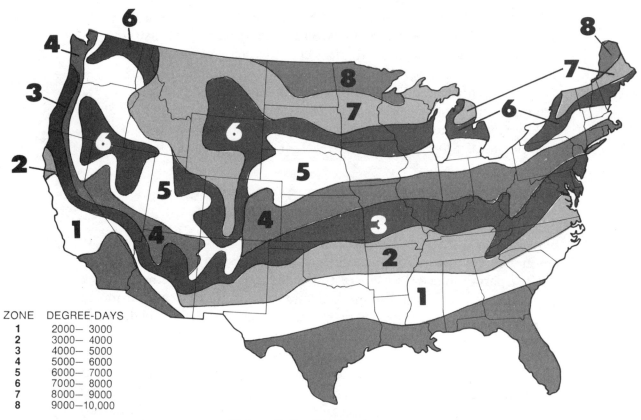

ZONE	DEGREE-DAYS
1	2000— 3000
2	3000— 4000
3	4000— 5000
4	5000— 6000
5	6000— 7000
6	7000— 8000
7	8000— 9000
8	9000—10,000

Figure 3.6 Thermal zones

Adapted from a National Oceanic and Atmospheric Administration/Environmental Data Service map. May not be accurate in mountainous regions. For more precise degree-day data, contact the nearest office of the National Weather Service.

CONDUCTIVE HEAT SAVING

(Heat units per square foot per year, by degree-day zone. Numbers in parentheses indicate amount of air space, in inches, between panes of glass.)

Zone 1

Current Windows	New Windows					
	Double-glazed (3/16)	Double-glazed (1/4)	Storm Windows (1 to 4)	Double-glazed (1/2)	Triple-glazed (1/4)	Triple-glazed (1/2)
Single glass	30	32	37	38	44	49
Double-glazed (3/16)		2	7	8	14	19
Double-glazed (1/4)			5	6	12	17
Storm windows (1 to 4)				1	7	12
Double-glazed (1/2)					6	11
Triple-glazed (1/4)						5

Zone 2

Current Windows	New Windows					
	Double-glazed (3/16)	Double-glazed (1/4)	Storm Windows (1 to 4)	Double-glazed (1/2)	Triple-glazed (1/4)	Triple-glazed (1/2)
Single glass	41	45	52	53	61	68
Double-glazed (3/16)		3	10	11	20	27
Double-glazed (1/4)			7	8	16	23
Storm windows (1 to 4)				1	10	16
Double-glazed (1/2)					9	16
Triple-glazed (1/4)						7

Zone 3

Current Windows	New Windows					
	Double-glazed ($\frac{3}{16}$)	Double-glazed ($\frac{1}{4}$)	Storm Windows (1 to 4)	Double-glazed ($\frac{1}{2}$)	Triple-glazed ($\frac{1}{4}$)	Triple-glazed ($\frac{1}{2}$)
Single glass	53	57	66	67	78	87
Double-glazed ($\frac{3}{16}$)		4	13	14	25	34
Double-glazed ($\frac{1}{4}$)			9	10	21	30
Storm windows (1 to 4)				1	12	21
Double-glazed ($\frac{1}{2}$)					11	20
Triple-glazed ($\frac{1}{4}$)						9

Zone 4

Current Windows	New Windows					
	Double-glazed ($\frac{3}{16}$)	Double-glazed ($\frac{1}{4}$)	Storm Windows (1 to 4)	Double-glazed ($\frac{1}{2}$)	Triple-glazed ($\frac{1}{4}$)	Triple-glazed ($\frac{1}{2}$)
Single glass	65	70	81	82	95	106
Double-glazed ($\frac{3}{16}$)		5	16	17	31	42
Double-glazed ($\frac{1}{4}$)			11	12	26	36
Storm windows (1 to 4)				1	15	26
Double-glazed ($\frac{1}{2}$)					13	24
Triple-glazed ($\frac{1}{4}$)						11

Zone 5

Current Windows	New Windows					
	Double-glazed ($\frac{3}{16}$)	Double-glazed ($\frac{1}{4}$)	Storm Windows (1 to 4)	Double-glazed ($\frac{1}{2}$)	Triple-glazed ($\frac{1}{4}$)	Triple-glazed ($\frac{1}{2}$)
Single glass	76	82	95	97	112	125
Double-glazed ($\frac{3}{16}$)		6	19	21	36	49
Double-glazed ($\frac{1}{4}$)			13	14	30	43
Storm windows (1 to 4)				2	17	30
Double-glazed ($\frac{1}{2}$)					16	29
Triple-glazed ($\frac{1}{4}$)						13

Zone 6

Current Windows	New Windows					
	Double-glazed (³⁄₁₆)	Double-glazed (¼)	Storm Windows (1 to 4)	Double-glazed (½)	Triple-glazed (¼)	Triple-glazed (½)
Single glass	88	95	109	111	130	144
Double-glazed (³⁄₁₆)		7	22	24	42	57
Double-glazed (¼)			15	16	35	49
Storm windows (1 to 4)				2	20	35
Double-glazed (½)					18	33
Triple-glazed (¼)						15

Zone 7

Current Windows	New Windows					
	Double-glazed (³⁄₁₆)	Double-glazed (¼)	Storm Windows (1 to 4)	Double-glazed (½)	Triple-glazed (¼)	Triple-glazed (½)
Single glass	99	107	124	126	147	163
Double-glazed (³⁄₁₆)		8	25	27	47	64
Double-glazed (¼)			17	19	39	56
Storm windows (1 to 4)				2	23	39
Double-glazed (½)					21	37
Triple-glazed (¼)						17

Zone 8

Current Windows	New Windows					
	Double-glazed (³⁄₁₆)	Double-glazed (¼)	Storm Windows (1 to 4)	Double-glazed (½)	Triple-glazed (¼)	Triple-glazed (½)
Single glass	111	120	138	141	164	182
Double-glazed (³⁄₁₆)		9	28	30	53	71
Double-glazed (¼)			18	21	44	62
Storm windows (1 to 4)				2	25	44
Double-glazed (½)					23	41
Triple-glazed (¼)						18

information for the Columbus area (zone 4) shows a saving of 81 heat units per square foot. In column 2, enter the heat units saved.

3. Multiply column 1 by column 2. Enter the result in column 3. Note: If you are replacing average or tight windows with thermal windows, skip steps 4, 5, and 6.

4. Measure the cracks around the windows. This will allow you to calculate the heat to be saved by reducing the amount of air leakage. The crack length for each window will be twice the height plus twice the width; if you have double-hung windows, include the width in the middle of the window, where the two sashes meet. Take all the measurements in inches, do the necessary addition, then divide by 12. Repeat for each window. (The example shows a total of 360 feet of crack.) Enter the total crack length (in feet) in column 4.

5. Determine the air-leakage saving. Find the air-leakage table for your zone. Move down the left-hand column to find the type of window that most closely matches what you now have. A *loose* window has a sash that rattles in high winds or when you shake it. A *tight* window is weatherstripped and doesn't rattle. An *average* window falls between the two extremes. Read across the table to find the type of window that most closely matches what you want to install. There you will find the amount of heat to be saved by reducing air leakage. If you are replacing loose windows with thermal windows, use the saving given for weatherstripping loose windows in your zone. (The saving is 24 heat units for the hypothetical house, assuming that the original windows were loose.) Enter the air-leakage saving in column 5.

6. Multiply column 4 by column 5. Enter the result in column 6.

7. Total the number of heat units saved. Add columns 3 and 6. Enter the total in column 7.

8. Enter the fuel factor. This number takes into account the different amounts of heat produced by different fuels. The factor for oil is .009; for natural gas, it's .013; for electricity, it's .29; for propane, it's .059 if you buy it by the pound, .013 if you buy it by the gallon. Enter in column 8 the fuel factor appropriate for your heating system.

9. Multiply column 7 by column 8. Enter the result in column 9.

10. Factor in the cost of fuel. Find out how much you pay for electricity, oil, natural gas, or propane—not the total monthly bill, but the rate for a basic unit of fuel: kilowatt-hour, gallon, or 100 cubic feet (about 1 therm). Your monthly statement should give the rate, but if it doesn't, call your utility company or fuel dealer. The hypothetical homeowner in our example pays 57 cents per 100 cubic feet for natural gas (entered as $0.57, not 57 cents).

11. Multiply column 9 by column 10. Enter the result in column 11. This figure will be your total saving, per heating season, if you add storm windows.

AIR-LEAKAGE SAVING

(Heat units per linear foot per year, by degree-day zone)

Zone 1

Condition of Current Windows	New Windows				
	Weather-stripping	Loose storm windows	Tight storm windows	Loose storm windows and weather-stripping	Tight storm windows and weather-stripping
Loose fit (no weather-stripping)	11	4	8	11	11
Average fit (no weather-stripping)	5	1	4	5	5
Tight fit (with weather-stripping)	0	0	1	1	1

Zone 2

Condition of Current Windows	New Windows				
	Weather-stripping	Loose storm windows	Tight storm windows	Loose storm windows and weather-stripping	Tight storm windows and weather-stripping
Loose fit (no weather-stripping)	15	5	12	15	15
Average fit (no weather-stripping)	7	1	6	7	8

Condition of Current Windows	New Windows				
	Weather-stripping	Loose storm windows	Tight storm windows	Loose storm windows and weather-stripping	Tight storm windows and weather-stripping
Loose fit (no weather-stripping)	15	5	12	15	15
Average fit (no weather-stripping)	7	1	6	7	8
Tight fit (with weather-stripping)	0	0	1	1	1

Zone 3

Condition of Current Windows	New Windows				
	Weather-stripping	Loose storm windows	Tight storm windows	Loose storm windows and weather-stripping	Tight storm windows and weather-stripping
Loose fit (no weather-stripping)	19	6	15	19	20
Average fit (no weather-stripping)	9	1	8	9	10
Tight fit (with weather-stripping)	0	0	1	1	1

Zone 4

Condition of Current Windows	New Windows				
	Weather-stripping	Loose storm windows	Tight storm windows	Loose storm windows and weather-stripping	Tight storm windows and weather-stripping
Loose fit (no weather-stripping)	23	8	18	23	24
Average fit (no weather-stripping)	11	2	9	11	12
Tight fit (with weather-stripping)	0	0	1	1	1

Zone 5

Condition of Current Windows	New Windows				
	Weather-stripping	Loose storm windows	Tight storm windows	Loose storm windows and weather-stripping	Tight storm windows and weather-stripping
Loose fit (no weather-stripping)	27	9	21	27	28
Average fit (no weather-stripping)	12	2	11	13	14
Tight fit (with weather-stripping)	0	0	1	1	1

Zone 6

Condition of Current Windows	New Windows				
	Weather-stripping	Loose storm windows	Tight storm windows	Loose storm windows and weather-stripping	Tight storm windows and weather-stripping
Loose fit (no weather-stripping)	31	11	25	31	32
Average fit (no weather-stripping)	14	2	13	15	16
Tight fit (with weather-stripping)	0	0	2	2	2

Zone 7

Condition of Current Windows	New Windows				
	Weather-stripping	Loose storm windows	Tight storm windows	Loose storm windows and weather-stripping	Tight storm windows and weather-stripping
Loose fit (no weather-stripping)	35	12	28	35	37
Average fit (no weather-stripping)	16	3	14	17	18
Tight fit (with weather-stripping)	0	0	2	2	2

Zone 8

Condition of Current Windows	New Windows				
	Weather-stripping	Loose storm windows	Tight storm windows	Loose storm windows and weather-stripping	Tight storm windows and weather-stripping
Loose fit (no weather-stripping)	39	13	31	39	41
Average fit (no weather-stripping)	18	3	16	19	20
Tight fit (with weather-stripping)	0	0	2	2	2

Calculating the Saving for Weatherstripping Alone

1. Disregard conductive saving and begin your calculations with step 4. Measure the total crack length for your windows, then determine the heat saved by reducing air-leakage losses. Enter the data in columns 4 and 5.

2. Multiple column 4 by column 5. Enter the result in column 6.

3. Enter the fuel factor in column 8, then multiply column 6 by column 8. Enter the result in column 9.

4. Enter the fuel cost in column 10. Multiply column 9 by column 10. Enter the result in column 11. The result will be the total saving in dollars per heating season that weatherstripping will provide. The calculations for the Columbus house are shown on the second line of the worksheet.

INSULATION

Of all the energy-saving products on the market today, insulation is apt to be the first one the homeowner considers. While there's no question that every home should be insulated adequately, insulation may have been oversold as

WINDOW SAVING CALCULATOR WORKSHEET

1 Total Window Area	2 Conductive Heat Saving per Square Foot	3 Total Conductive Saving	4 Total Crack Length (in Feet)	5 Air-leakage Saving per Foot	6 Total Air-leakage Saving	7 Overall Heat Saving (Col. 3 + Col. 6)	8 Fuel Factor	9 Fuel Units Saved	10 Fuel Cost per Fuel Unit	11 Annual Saving
280 × 81		=22,680	360 × 24		=8,640	31,320 ×	.013	=407.16 ×	$0.57	=$232.08
___ × ___		=___	360 × 23		=8,280	8,280 ×	.013	=107.64 ×	$0.57	=$61.35
___ × ___		=___	___ × ___		=___	___ ×	___	=___ ×	___	=$___
___ × ___		=___	___ × ___		=___	___ ×	___	=___ ×	___	=$___

a conservation tool. If your house already has some insulation, adding more may not be the best course to follow: The money invested often will yield a surprisingly small saving of energy and money. One government study found that adding more insulation to the attic of a test home yielded little saving—only 6 percent. (The biggest saving came from the installation of storm windows.)

What's more, installing insulation has certain drawbacks. Blowing it inside walls or other finished areas can be chancy, since there's no way to be sure that the area has been filled properly. Faulty installation can also lead to moisture condensation within walls, ceilings, and floors, and loose-fill types may settle in time.

Why Insulate?

Nearly any material will slow the flow of heat, but some materials do a better job than others. Six inches of fiberglass insulation, for instance, is as effective as a brick wall more than eight feet thick. As noted earlier in this book, the measure of a material's ability to retard heat flow is known as the R-value; the higher the R-value of a given material, the better it is as an insulator. Therefore, in estimating your home's insulation needs, think only in terms of R-value, not in terms of a specific type of material or in terms of inches of insulation. It's the R-value that matters most. Remember that 4½ inches of insulation rated at R-4 will be as effective as 6 inches of insulation rated at R-3 (4½ × 4 and 6 × 3 both equal 18).

Because R-values are additive, two fiberglass batts, each rated R-19, together yield R-38. The siding, roofing and flooring in a house also have R-values of their own, albeit small ones. An uninsulated attic floor isn't R-0; it's normally anywhere from R-2 to R-4.

A little more than a generation ago, when oil and gas were cheap, most homes were built with little or no insulation. It was no great economic hardship to let the furnace run and keep the house comfortable. But ever since the cost of home heating began to rise, insulation has become increasingly important.

When Enough Is Enough

It's important to keep in mind that you *can* insulate a house to a point where the money spent far exceeds the money saved on fuel. The Insulation Worksheet in this chapter tells you how to figure out where that point comes. It may seem like a fair amount of work to use, but once you have done the estimating we recommend, you'll be able to judge where to spend your money wisely—whether for extra attic insulation, new insulation in walls or crawl spaces, or some other kind of energy-saving product.

Suppose, for example, that a homeowner in Columbus, Ohio, wants to insulate the attic. Assume that the attic has 666 square feet of floor area, no insulation, and that the house is heated with natural gas costing 56 cents per 100 cubic feet. According to the calculations, installing R-19 insulation would save about $132 per heating season.

The saving would be much less if this hypothetical Ohio home already had some attic insulation. With, say, R-11 in place, an additional R-19 (for a total R-value of 30) would save only about $40 more per heating season. That kind of slight annual saving might lead the homeowner to consider leaving the attic at R-11 and insulating the walls instead.

This time, suppose the house has 1,600 square feet of uninsulated wood-frame walls with wood siding. The walls now have an R-value of 4. By hiring a contractor to blow in fiberglass, the homeowner could bring the walls to R-12. The annual saving: slightly more than $130.

These examples are *only* for the heating season, of course. For a discussion of insulation's effect on air-conditioning requirements, see chapter 7.

How quickly should an investment in insulation pay for itself? No single number of months or years is right for everyone. But by calculating the saving for different amounts of insulation, or for different areas of the house, you will learn where added insulation could bring the greatest saving—and where the saving will mount up most quickly.

What about the effect of inflation and other economic facts of life? Regardless of what happens, it still makes sense to put your energy-conservation dollars where they will do the most good *now*. The best investment today will also be the best investment in the future, regardless of higher fuel prices or higher material costs.

If your calculations show that it will take considerable time to save enough on fuel to recover the cost of insulation, you should consider investing in other energy-savers instead. The most effective of these include storm windows, exterior caulking, and weatherstripping to eliminate drafts around doors and windows; and an automatic-setback thermostat. The worksheets in this book can help you make a sound decision.

Types of Insulation

Your calculations may show that additional insulation is indeed a wise choice. If so, what kind of insulation will work best? It's more important to understand the differences among types of insulation than to recognize brand names. There are several varieties, each with its own advantages and disadvantages. They include the following (all R-values are per inch of installed thickness):

Batts and blankets: fiberglass or mineral wool (R-3.1). Suitable for use in unfinished areas, batts and blankets can be do-it-yourself items. They are made in standard thicknesses (generally yielding total R-values of 11, 13, 19, and 22) and in widths meant to fit between wall studs and floor joists. They are available with or without a vapor barrier. A barrier is necessary for most types of insulation (not just fiberglass) to minimize moisture condensation in walls, floors, and ceilings. The risk of fire from fiberglass and mineral wool is very remote.

Bulk insulation/loose-fill: cellulose (R-3.7), mineral wool (R-3.1), perlite (R-2.6), fiberglass (R-2.3), and vermiculite (R-2.1). These materials are suitable for use in both finished and unfinished areas. They can be poured into place between exposed floor joists or blown into finished wall or floor cavities. Pouring insulation is a job most homeowners should be able to handle themselves; blowing in insulation is a job for an experienced contractor. Any loose-fill material may settle in time, thus lowering the R-value. Except for improperly treated or badly installed cellulose, the risk of fire from loose-fill is remote.

Plastic foam boards: urethane (R-5 to R-8), polystyrene (R-4.5), and bead-board (R-3.6). These products are suitable for use as exterior sheathing, or to cover finished walls. To insulate finished walls (something you can handle if you're a fairly accomplished do-it-yourselfer), panels of material are attached to an existing wall, then covered with a vapor barrier and, for fire protection, gypsum board at least ½ inch thick. Plastic boards don't shrink or settle, but they are combustible. They *must* be covered with a fire-retardant material, as specified by your local building code.

Bulk insulation/foam: urea-formaldehyde (R-4.8). The hazards of form-aldehyde in the home came to public attention some years ago largely because of its presence in urea-formaldehyde foam insulation. Many home-owners had this foam pumped into their hollow exterior walls to save on energy costs—only to find that formaldehyde gas from incorrectly prepared or applied foam could cause eye and respiratory irritation, sometimes severe. Beyond those immediate health problems, an industry-sponsored study showed that long-term high exposures to formaldehyde caused nasal cancer in rats.

The Consumer Product Safety Commission (CPSC) banned the use of urea-formaldehyde foam insulation in 1982. A federal appeals court over-turned the ban the next year, but even before the ban took effect, the adverse publicity had basically halted the use of the foam. Nevertheless, it is still avail-able. Prudence suggests that it be avoided. Beyond the health questions, the foam can shrink after installation, drastically reducing its insulating effec-tiveness. The material also will burn, although it presents little risk of fire inside walls.

Installation

No matter which type of insulation you use, or who does the installing, here are five points to keep in mind:

1. It's important to include a vapor barrier (a sheet of foil, plastic, or treated paper) facing the inside of the house to minimize moisture condensation, which could soak insulation and lower the R-value. Moisture might also cause studs, joists, and sheathing to rot, and exterior paint to blister. If you are insulating finished walls, apply two coats of oil-based or aluminum paint

to the inside walls for some protection against condensation.

2. Don't block vents in eaves or crawl spaces with insulation; they are important for temperature and moisture control.

3. Keep insulation away from the tops and sides of recessed lighting fixtures and o...er heat sources. (If you are planning to use loose-fill, you'll need to install a shield of some sort to keep the material in place.)

4. If you are handling the insulation, wear protective clothing—gloves, long-sleeved shirt, long trousers, a dust mask, and goggles—to keep dust or stray mineral-wool fibers away from your skin, lungs, and eyes.

5. The R-values given here are those generally accepted by both government and industry. Don't deal with contractors or suppliers who claim markedly higher R-values for these materials. And be sure to check the label; it should carry the following statement: "R means resistance to heat flow. The higher the R-value, the greater the insulating power." The label should also include a chart showing such informaton as the length, width, and thickness of the insulation in the package. If the label does not provide such information, don't buy the product.

Keep in mind that costs and installation problems will vary from one part of the house to another. Unfinished areas—attic floors, crawl spaces, basement walls or ceilings—are the easiest to get to and generally the least expensive to insulate. Finished attic floors and outside walls are the most difficult and the costliest. In fact, as we have noted, it may not be wise to reinsulate walls that already have some insulation.

It's not necessary to reinsulate the entire house at once. If you find that your attic needs insulation, but your home-improvement budget will allow you to cover only half the area, fine; you will most likely save enough on fuel to recover that cost quickly. You can insulate the rest of the attic later.

Prices for insulation vary widely from one area of the country to another. If you are planning to install insulation yourself, shop for the best price. Otherwise, shop for the best *contractor,* even if that means paying a premium, to improve the chances of getting insulation installed safely and properly (see chapter 9).

INSULATION WORKSHEET

This worksheet can help you compare the saving you could achieve by insulating your house in various ways with storm windows, insulated shutters, or the other heat-savers for windows described in this chapter; or with insulation in the walls, ceilings, or floors. You will then be able to determine where insu-

1. Unfinished attic; crawl space (not shown).

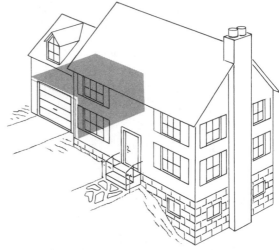

2. Walls and ceilings between unheated parts of the house (such as a garage) and living area.

3. Ceiling in unfinished, unheated basement.

4. Exterior walls, including finished attic (left) and finished basement (not shown).

In theory, every building surface that separates the living area from the outside should be insulated. However, you don't have to insulate the entire house at once. Start with the areas that are the easiest and least expensive, then do the more difficult and more expensive areas. Follow the general sequence indicated here.

Figure 3.7   Where to insulate

lation would yield the greatest saving. To guide you through the worksheet, there is the example of a homeowner in Columbus, Ohio, who is weighing the saving offered by storm windows against that of extra insulation in the attic. First, you must calculate a "fuel/climate cost factor" for your area. That done, you can begin to work out your saving in lines 6 to 10. (If you use a pocket calculator, you can do all the arithmetic in one step instead of stopping to enter subtotals.)

I. Fuel/Climate Cost Factor

Enter the cost of fuel. Enter on line 1 the price you now pay for electricity, oil, natural gas, or propane—not your total monthly bill, but the amount you pay per kilowatt-hour, gallon, or 100 cubic feet. (Our hypothetical home-owner uses natural gas at 56 cents per 100 cubic feet.)

Enter the fuel factor. This number takes into account the different amounts of heat produced by different fuels. The factor for oil is .009; for natural gas, .013; for electricity, .29; and for propane, .059 if the fuel is purchased by the pound, .013 if purchased by the gallon. Choose the appropriate factor and enter it on line 2. (The fuel factor is .009 in our example.)

Multiply line 1 by line 2. Enter the result on line 3.

Determine the climate factor for your area. The climate factor takes into account the average differences between indoor and outdoor temperatures in various parts of the country over the course of a heating season; in effect, the factor acknowledges that it's colder in the north than in the south. To deter-mine the correct factor, consult the map included with this worksheet to find the zone in which you live. Then find the climate factor for that zone in the table accompanying the map. Enter your climate factor on line 4. (Our hypo-thetical homeowner lives in zone 4, which has a climate factor of 134.)

Multiply line 3 by line 4. Enter the result on line 5. The number is your fuel/climate cost factor. (The factor for our hypothetical homeowner is 1.495.)

II. Costs and Savings

Enter your fuel/climate cost factor (line 5) in the spaces on line 6, as shown in our example. (For our hypothetical homeowner, who wants to compare the saving from storm windows against the saving from extra insulation, the number has been entered in all four spaces.)

Enter appropriate R-values. Of all the steps on the worksheet, this is the most time-consuming. You'll need to figure the present R-value for the wall, floor, ceiling, or window that you want to insulate. If your house already has some insulation, figuring the present R-value will involve some addition—the R-value for the structure itself plus the R-value for the existing insulation. You'll also need to figure the appropriate new R-values for the various options—adding the R-value of the new insulation to the present R-value. The accompanying table of R-values gives values for some typical kinds of construction and for the commonly used kinds of insulation. The example should make this step clear. The hypothetical homeowner now has single-glazed windows, so 0.9 is entered in the first space on line 7. The homeowner wants to add storm windows, which have an R-value of 1.1. Once the storm windows are up, the windows will have an R-value of 2 (0.9 for the existing windows plus 1.1 for the storm windows). So, 2 is the figure entered in the second space on line 7. For the existing attic insulation, the homeowner uses a figure of 14. That is the sum of 3 for the ceiling itself, plus 11 (the R-value for 3½ inches of fiberglass insulation already in place). The number 14 is entered in the third space on line 7. Adding another 3½ inches of fiberglass insulation will increase the total R-value of the attic to 25 (14 plus 11). Calculate the present R-values for the areas you want to insulate and enter them in the appropriate spaces; work out the new R-values and enter them.

Divide line 6 by line 7. Enter the result on line 8. This will determine the cost of heat lost per square foot each year—both with your house as it now exists and after the improvements you have planned. (In the example, the homeowner is now paying $1.09 for the heat lost through each square foot of window. Adding storm windows would reduce that loss to 49 cents per square foot. For insulation, the present cost of heat loss per square foot, rounded off, is about 7 cents. Added insulation would reduce that to 4 cents per square foot.)

Determine the area you wish to insulate. Measure each window's height and width, in feet; multiply those two numbers to determine the area for each window, then add the areas of all the windows you wish to insulate. For walls, floors, and ceilings, measure the width and length, in feet, and be sure to subtract the areas of doors and windows. Because there will be studs or floor joists in the way, multiply the overall square footage by 0.9 to compensate for the area they occupy. Once you have determined the total area to be insulated, enter the numbers in the appropriate spaces on line 9. (The hypothetical homeowner's attic area measures 1,670 square feet, for a net area of about 1,500 square feet—1,670 times 0.9. Each window is 15 square feet, and 10 windows are to be insulated, for an overall window area of 150 square feet.)

Multiply line 8 by line 9. Enter the result on line 10. In the columns for present windows and insulation, the answer will be the amount you now spend each year for heat lost through the walls, floors, or windows. The columns for improved windows and insulation indicate the reduced costs after improvements are made. By comapring present-condition costs with improved-condition costs, you can see what your saving is likely to be. (The Columbus homeowner stands to have a saving of about $90—$164 minus $74—by installing storm windows. The saving from added insulation would be about $45—$105 minus $60.)

INSULATION WORKSHEET

I. Fuel/Climate Cost Factor

	Example	Your House
1. Fuel cost (price per kwh, gallon, or 100 cu. ft.)	$ .56	$ _____
2. Fuel factor (see instructions)	× .013	_____
3. Subtotal (line 1 × line 2)	= .0073	_____
4. Climate factor (from table at right)	× 134	_____
5. Fuel/climate cost factor (line 3 × line 4)	= $ .98	_____

II. Costs and Savings

	Example				Your House			
	Present Windows	Improved Windows	Present Insulation	Improved Insulation	Present Windows	Improved Windows	Present Insulation	Improved Insulation
6. Fuel/climate cost factor (from line 5)	$ .98	$ .98	$ .98	$ .98	$ ___	$ ___	$ ___	$ ___
7. R-value (see instructions and table)	÷ 0.9	2	14	25	___	___	___	___
8. Cost of heat loss per square foot (line 6 ÷ line 7)	= $1.09	$ .49	$ .07	$ .04	$ ___	$ ___	$ ___	$ ___

	Example				Your House			
	Present Windows	Improved Windows	Present Insulation	Improved Insulation	Present Windows	Improved Windows	Present Insulation	Improved Insulation
9. Area to be insulated (see instructions)	× 150	150	1500	1500	_____	_____	_____	_____
10. Cost of heat loss (line 8 × line 9)	= $ 164	$174	$105	$160	$____	$____	$____	$____

R-Values for Building Materials and Insulation

Buildings		Insulation	
Walls		**Batts or blankets,**	
Uninsulated wood frame, with		**fiberglass**	
Stucco	R-3	**or mineral wool**	R-3.1/in.
Asbestos siding	R-3	**Bulk/loose-fill**	
Wood siding	R-4	Cellulose	R-3.7/in.
Face-brick veneer	R-4	Mineral wool	R-3.1/in.
Wood shingles	R-5	Perlite	R-2.6/in.
Brick	R-4	Fiberglass	R-2.3/in.
Concrete block	R-4	Vermiculite	R-2.1/in.
Ceilings (under attics)		**Plastic foam boards**	
Plaster or plasterboard	R-2	Urethane	R-5 to R-8/in.*
Acoustical tile	R-3	Polystyrene	R-4.5/in.
(add 2 if attic has subfloor)		Beadboard	R-3.6/in.
Wood floors over basements, porches, and crawl spaces (add 2 for finished ceiling in basement)	R-3		
Windows			
Single-glazed	R-0.9		
Storm window (interior or exterior)	R-1.1		

*Check labeled R-value for particular product

INDOOR AIR POLLUTION:
CAN A HOUSE BE SEALED TOO WELL?

Indoor air can be dirtier than outdoor air. For some pollutants, indoor exposures regularly exceed permissible outdoor standards set for them. Since most people spend more than 60 percent of their time inside their homes, scientists are concerned that high indoor pollutant levels might have adverse health effects.

Two factors have aggravated the indoor-air problem: New chemical products have found their way into houses, and those houses are better insulated than they used to be. Concern has also increased as reseachers have become more knowledgeable about chemical hazards that have long existed in people's homes.

Thousands of chemicals—some toxic and many never adequately tested for toxicity—are present in household products ranging from paint strippers to insecticides. Additional chemicals diffuse into the indoor air from furnishings and building materials.

The "tighter" the house, the more serious the problem can be. Caulking, weatherstripping, and insulation cut down on ventilation and thus help seal in pollutants. Most susceptible to indoor pollution problems are the so-called supertight houses that have been built since the mid-1970s.

In the typical American home, an amount of fresh air equal to the volume of air in the house leaks in about once an hour. That's called a ventilation rate of 1.0 air change per hour. But supertight houses may have ventilation rates of only 0.1 or 0.2 air change per hour.

Ordinary tightening of an existing house generally results in a far smaller reduction in ventilation rates. Typical measures—caulking, weatherstripping, increased insulation—reduce air circulation by 25 percent at most. Unless a powerful pollutant is present, this ordinarily would not push indoor pollution concentrations to extremely high levels. Nevertheless, you should be aware of the signs of possible pollution problems. These include condensation on the insides of windows in winter; the presence of mold or mildew on walls or ceilings; stale odors that linger; smarting eyes; or frequent respiratory illnesses, especially among children. Improved ventilation often is the solution to these symptoms of indoor pollution.

The pollutants that pose the greatest threats inside people's homes are not necessarily the same ones that pose the biggest problems outdoors.

Household Chemicals

A typical house harbors dozens of products that release organic chemicals. Many household-chemical products—spray paints, insecticides, furniture polish, and so on—come in aerosol form, ensuring that tiny droplets of the product will be dispersed into the air and adding an additional chemical (the propellant) to those that are present in the basic product. Some of these chemicals, particularly solvents, may pose hazards to your health.

Radon

Scientists have long puzzled over what causes the roughly 15 percent of lung-cancer cases that occur in people who don't smoke. Increasing evidence suggests that the second leading cause of lung cancer may well be exposure to radon gas.

Of all the indoor pollutants we encounter, radon is probably the most dangerous. This naturally occurring radioactive gas can be found under the earth in varying amounts virtually everywhere in the world. Like any gas, it diffuses out of the ground and into the air—or into the houses that happen to be built above. Houses can trap radon gas that otherwise would disperse into the atmosphere.

The only way to find out whether you have a radon problem is to have a radon measurement. You might want to check with your state or local health department to see if a testing program is available in your area. Consumers Union tested a few do-it-yourself detectors, the kind that you return to their suppliers for analysis after a suitable period of exposure in your home. They all provided at least a rough approximation of radon levels in the test house.

Formaldehyde

There are many sources of formaldehyde other than from the blown-in house insulation discussed earlier. Each year, billions of pounds of it are used in making plywood and particleboard and in treating textiles such as those used for many draperies and carpets. Mobile homes, which are tightly constructed and manufactured with a great deal of particleboard, are especially prone to indoor formaldehyde problems.

While formaldehyde is not odorless, the gas can have irritant effects at levels that you can't smell. If you think you may have a formaldehyde problem in your house, formaldehyde monitors are available at moderate cost. One company that offers them is Enys, Inc., P.O. Box 14063, Research Triangle Park, North Carolina 27709. Each detector is a small glass vial that you expose to room air for a week and then mail back to the company for analysis. The cost for a kit containing a pair of the detectors is $59, including postage, laboratory analysis, a written report, and a booklet explaining the results.

Combustion Products

Three of the major outdoor pollutants—nitrogen dioxide, carbon monoxide, and particulate matter—are commonly found indoors at levels higher than those outdoors. Indoor exposures are often high enough to pose a hazard to human health. Their major sources are gas ranges, heating appliances (defective central heating systems, unvented gas and kerosene space heaters, wood-burning stoves), and cigarette smoke.

Nitrogen dioxide is an irritant gas that affects the respiratory tract. Long-term exposure to levels above the outdoor standards may contribute to respiratory disease. People who already have respiratory problems such as bronchitis and asthma are at particular risk from nitrogen-dioxide exposure.

Carbon monoxide—odorless, colorless, and undetectable to the senses—interferes with the blood's ability to carry oxygen to the cells of the body. People with angina pectoris (a heart condition characterized by chest pain) are among those most sensitive to carbon monoxide. Also sensitive to it are fetuses, newborns, and people with chronic lung disease or anemia.

Particulate matter includes a wide variety of substances that float in the air as discrete particles, either as solids or as liquid droplets. Particles may be toxic themselves or act as carriers for other toxic substances. Most harmful are small particles that, when inhaled, are carried deep into the lungs.

People at special risk from particulates include those with emphysema, bronchitis, asthma, or heart disease, plus smokers, children, and the elderly.

Here's a rundown on the major sources of combustion pollutants:

Gas ranges. If you own a gas range, a range hood vented to the outside effectively removes much of the nitrogen dioxide as well as the other pollutants that cooking creates. A kitchen exhaust fan can also be useful. Even an open window helps.

Unvented range hoods don't help much. Their filters may trap grease and odors, but they have no effect on gases, are expensive, and generally are not changed often enough to be useful. In addition, unvented hoods circulate exhaust fumes back toward the user's face.

Improperly adjusted burner flames can multiply the pollutant level. Be sure to adjust flames so that they burn blue instead of orange.

When you replace your old gas range, buy a model that doesn't use a pilot light. That will help cut down on pollutant emissions and also reduce your gas bill.

Gas space heaters. Unvented gas space heaters are found most commonly in rural areas of the South and the Southwest. Their high outputs of carbon monoxide and nitrogen dioxide probably make them the most polluting of all unvented combustion appliances.

In a mid-1980s study, researchers at the University of California's Lawrence Berkeley Laboratory measured pollutant levels produced by a variety of gas space heaters. For carbon monoxide, maximum levels measured indoors were nearly three times the long-term outdoor-air standard. Nitrogen-dioxide levels ranged from 8 to 29 times higher than the outdoor nitrogen-dioxide standard.

Tobacco smoke. Smoke from tobacco adds copious amounts of particulates to indoor air; it also adds hundreds of different gases, including carbon monoxide and nitrogen dioxide.

The impact of cigarettes on smokers is all too clear: Cigarettes account for a high percentage of the lung cancer deaths that occur each year in the United States. They also contribute to thousands of cases of emphysema, cardiovascular disease, and other ailments.

Cigarette smoke also harms certain "passive smokers"—people who inhale the smoke at home, at work, and elsewhere. Very young children of smoking parents have an increased incidence of bronchitis and pneumonia, and they are more likely to be hospitalized for respiratory infections than the children of nonsmokers. Many adults, especially people with respiratory allergies, suffer severe discomfort from the irritative effects of cigarette smoke. And the pregnancies of women who smoke involve a much-greater-than-normal risk of miscarriage and stillbirth. Babies born to smokers tend to have lower birthweight and an increased incidence of health and developmental problems after birth.

Since tobacco smoke inhaled by passive smokers contains the same carcinogens as the smoke inhaled by active smokers, it is believed that passive smoking poses a risk of lung cancer.

If you don't smoke yourself, you're certainly well within your rights to request that visitors to your home refrain from smoking. If someone in your home does smoke, consider purchasing an effective air cleaner.

Filters placed in forced-air heating systems or central air-conditioning systems can also help remove smoke particles. Most such systems come equipped with coarse, "low-efficiency" filters that remove large particles. To remove smaller inhalable particles, you can add on medium- or high-efficiency filters. These filters, however, tend to be expensive.

Instead of adding filters to your central heating or air-conditioning system, you may want to install an electrostatic-precipitator air cleaner. These devices impart an electric charge to particles, which then are attracted to a collecting plate. According to the U.S. Department of Energy, electrostatic precipitators compare favorably with medium- and high-efficiency filters. Their main advantage is that they don't impede air flow the way filters do. But their collecting plates must be replaced or cleaned every few months.

Recommendations

If possible, eliminate the source of pollutants. That would include tuning up all fuel-fired equipment (space heaters, gas ranges) to reduce the emission to the room of incomplete combustion products; cleaning vaporizers and humidifiers regularly and often to reduce the possibility that they may become a breeding ground for infectious microorganisms; and trying to reduce the amount of smoking in the house. Beyond these steps, the major ways of reducing indoor air pollution are by increasing ventilation and by cleaning the air.

Exhaust fans. When pollution is produced in a specific place, exhaust fans can help remove it and prevent its spread to the rest of the house. Fans are particularly useful in the kitchen, the site of much indoor air pollution. An exhaust fan built into a hood over the stove is the best type, but it must vent to the outdoors. The fan should be used whenever food is cooked, particularly if the cooking is done on a gas stove.

Windows. An open window is an obvious and effective means of ventilation—especially in mild weather when little heat will be lost. In the kitchen or bathroom, an open window is a good supplement to an exhaust fan. It's also a good idea to open windows when using highly volatile substances such as paints.

Dehumidifiers. These are effective in reducing excessive moisture and the problems associated with it, including fungus (mold and mildew) growth. But dehumidifiers must be kept clean to prevent bacteria from accumulating.

Filters. Good filters can be quite helpful in removing particulate matter from the air. However, the typical filter found in a household appliance is of limited value in this respect. Mesh filters in range hoods remove grease, and fiber filters used in air conditioners and furnaces remove large particles. Such filters generally cannot catch smaller particles, however, which are more harmful when inhaled.

FOUR

Central Heating Systems

Even if your house is relatively free from air leaks and adequately insulated, you can still save on fuel by improving the efficiency of your house's oil or gas heating system.

Lower the thermostat setting. By turning down the thermostat even slightly, you can reduce the amount of heat you use in your house. The smaller the difference between the temperature inside the house and the temperature outside, the more slowly the heat will escape. Lowering the thermostat setting reduces the temperature difference, thus slowing heat loss. (If you have electric heat, lowering the thermostat setting is in effect your only option. Fortunately, it's a step that can yield a worthwhile saving.)

Get as much heat as possible from the fuel when the furnace is on. Even the most efficient oil or gas furnace that you are apt to encounter loses 15 to 25 percent of the energy contained in the fuel it burns when it's operating. Naturally, these on-time losses are greater for a less efficient furnace. Having the heating system (including the flue and chimney) cleaned and serviced yearly can improve the furnace's heat output. Some of the other steps you can take to improve the furnace's efficiency, including information about high-efficiency furnaces and oil burners, are described later in this chapter.

Reduce the loss of heat when the furnace is off. Heat will escape up the flue even after the furnace shuts down. These off-time losses will continue as long as the temperature inside the furnace is higher than the outside temperature at the top of the chimney. In the course of a heating season, off-time losses can almost equal on-time losses. This chapter will explain how to save fuel by derating the furnace and also help you decide whether to install a flue damper.

Switch fuels. Gas has almost always been cheaper than oil, Btu for Btu. Switching from oil to gas might have some obvious advantages, at least for some people. A section later in this chapter discusses the possibilities.

Improve existing equipment. Allow free air movement around vents, radiators, or baseboard heaters by avoiding blocked openings from furniture that is too close or draperies that hang over a hot-air register.

Insulate hot pipes that pass through such unheated areas as attics or crawl spaces. Before insulating, however, seal any pipe leaks.

If the heating system circulates hot water through radiators, bleed any accumulated air from hot-water radiators before each heating season (and once or twice during the season). Open the small valve at the top of the radiator and use a container to catch the water that will flow from the valve after the air has been eliminated. Be careful, because the water may be hot, depending on whether or not the system is in operation.

Newer systems often have self-purging valves that make bleeding unnecessary; if you don't hear gurgling when the system starts circulating, you can assume it's free of air.

Many older systems use steam radiators. Check to be sure that each radiator's air valve is working properly. The valve should hiss slightly (or fairly loudly, depending on the design) when hot steam fills the radiator and pushes out the air. If the radiator becomes hot and no steam emerges from the valve, it is working properly. If there is steam, or if the radiator doesn't get hot, replace the valve.

Clean or change the filters in a hot-air system several times during the heating season. Warm-air registers should be kept unobstructed and free of accumulated dust and debris. Ducts that feed the registers should be insulated if they pass through unheated spaces on the way to occupied, heated areas.

Should you buy new equipment? This chapter provides information about new heating equipment—something to consider when it is no longer practical to alter an old system.

You may want to obtain a copy of the Gas Appliance Manufacturers Association's *Consumer Directory of Certified Efficiency Ratings* for up-to-date technical information about such matters as the heating capacity of brands and models for the following kinds of residential heating and water-heating equipment:

- gas central furnaces
- oil central furnaces
- gas boilers
- oil boilers
- gas room heaters
- gas floor furnaces
- gas wall furnaces
- gas water heaters
- oil water heaters
- electric water heaters, including the heat-pump type

The directory is published twice a year, in April and October, and may be available at your local public library. You can obtain a copy for $5 from GAMA Efficiency Certification Program, ETL Testing Laboratories, Inc., Industrial Park, Route 11, Cortland, New York 13045.

SAVING WITH A THERMOSTAT

The simplest way to lower your heating bills is to turn down the thermostat. Begin by lowering your normal daytime thermostat setting to the coolest comfortable temperature. And if you set back the temperature farther at night or when the house is empty, you will add to your saving.

No one can say precisely how much you'll save by setting back your thermostat. But, in general, the greater the setback and the longer the setback period, the greater your saving will be. Both the kind of heating system you have and the weather conditions in your locality will affect the actual amount of money you save.

A computer simulation prepared at the Oak Ridge National Laboratory showed the approximate saving that could be expected in a house that switched from a constant temperature of 68°F to two kinds of nighttime setbacks: from 68° during the day to 60° between 10 P.M. and 6 A.M., and from 68° to 55° in the same time period.

If, for example, there are about 4,000 heating degree-days in your area, and you use a 68°-to-60°F setback, you could save roughly $17 on every $100 you now spend for heating your home. And if you use a 68°-to-55° setback, you might expect to save $22 per $100 of your heating costs.

SAVING PER $100 OF PRESENT FUEL BILL (NIGHTTIME SETBACK)

Degree-days	68° to 60°F	68° to 55°F
2,000	$23	$30
3,000	20	25
4,000	17	22
5,000	14	20
6,000	12	18
7,000	10	16
8,000	9	14

As the table above shows, the greater the setback on your thermostat, the greater the saving. The warmer your locality, the higher the *percentage* of saving, though probably not the *total* saving. Don't be misled by the "saving per $100" figure. Homeowners who face cold winters run up very high fuel bills, so their overall dollar saving will probably be higher than for people living in warmer climates. In other words, it's more helpful for people in Chicago to save 20 percent of their fuel bill than for people in Los Angeles to save 80 percent of theirs.

Automatic-Setback Thermostats

By lowering a conventional thermostat manually, you can save money without any additional expense. But you may find that turning down a thermostat at night is a chore easily forgotten. And you may not like getting out of a warm bed in a cold house to turn the thermostat back up in the morning and then having to wait for the house to get warm. Nevertheless, forgetting to set back your thermostat at night means you are passing up a chance to save energy and money.

You can do away with the nuisance and achieve a regular, consistent saving on fuel by installing a thermostat that controls to at least two temperatures so that the heat is automatically turned down at night and up again before you get out of bed the next morning. (You may find that when you no longer get up to a cold house, you'll set your night temperature even lower.) Whether it's worthwhile for you to purchase such a thermostat depends on several factors: where you live, how long it takes for your home to warm up, whether you're likely to forget to lower the thermostat at night, and personal preference.

Consumers Union has tested automatic-setback thermostats several times.

Look for *two-wire* models, which need no new wiring to replace an existing manual thermostat. (Some thermostats, particularly those that control central air conditioning as well as heating, require four or more connections.) Any reasonably handy person should be able to install a two-wire unit.

Clock-timer thermostats have a timer dial with movable pins that switch the unit between normal and setback temperatures. By moving the pins, you can control the duration of the setback periods. You then set the units for the normal and setback temperatures desired.

Mechanical thermostats have to be wound, like a kitchen timer, to start each setback period and determine its duration. These models are best suited for households that don't keep to a regular schedule.

In Consumers Union's most recent tests, a Honeywell T8082A (list price, late 1990, about $100) ranked highest among the automatic-setback units.

Fooling a Thermostat

One *Consumer Reports* reader wrote to describe a simple, cheap way to obtain the advantages of an automatic thermostat from a manual unit. Attach an ordinary night-light to the wall, about an inch below the thermostat, and plug the night-light into an ordinary timer. Set the timer to turn on the light when you want the heat reduced and to turn off the light when you want the heat raised. Responding to the bulb's heat, the thermostat would turn the furnace off and on. This would require some experimentation, involving moving the light a number of times, to get the desired amount of temperature setback.

This setup should work, but it (like some ready-made "thermostat foolers") is much less convenient to adjust than an automatic thermostat.

IMPROVING YOUR FURNACE'S EFFICIENCY

In homes with electric heat, the system's efficiency is so high that there is little or nothing you can do to improve it. Furnaces fueled with oil or gas are a different story. With these systems, you can take some steps to improve the furnace's performance when it's running. (Later in the chapter, we describe some steps to reduce heat loss after the furnace shuts off.)

Testing Combustion Efficiency

Combustion efficiency (sometimes referred to as CE) is the term commonly used to describe how good a job a furnace does when it's running—its on-time performance. Expressed in percentages, combustion efficiency equals the portion of the total energy in the fuel that becomes usable heat. A gas-fired or oil-fired furnace in good working order may have a combustion efficiency of roughly 75 or 80 percent, respectively. That is, for every 100,000 Btu of fuel entering the system when the furnace is on, only 75,000 or 80,000 Btu stay within the house as usable heat. And that's with quite efficient furnaces; many furnaces do considerably worse. For the most part, the heat that's lost goes up the chimney.

The first step toward trimming on-time losses is to have the furnace's combustion efficiency checked. Chances are the CE is somewhat below the best possible level. If the furnace hasn't been serviced regularly, internal soot and dirt could be reducing the flow of heat. The burner's nozzle or a gas orifice might need to be cleaned and replaced, air inlets and controls adjusted, and filters changed. You can have your heating-oil company or the gas company check combustion efficiency by measuring the temperature, smoke level, and carbon-dioxide content of the gases entering the flue.

Actually, if you are an experienced do-it-yourselfer, you can determine your furnace's combustion efficiency on your own—provided you're willing to spend about $385 for the required instruments. (You could share the cost with neighbors.) Here are the things you'll need:

A Tempoint thermometer (catalog no. 12-7014) to measure the temperature in the flue would cost about $42.50; a True-Spot Smoke Tester (catalog no. 21-7006) would cost abut $69; a Fyrite CO_2 indicator (catalog no. 10-5000) would run about $177; and an MZF draft gauge (catalog no. 13-7019) is priced at $139. These items are available as a kit (catalog no. 10-5022) for $385 (all prices as of late 1990). The manufacturer provides a calculator similar to a slide rule to convert the measurements to combustion efficiency. The equipment is sold (through their distributors) by Bacharach Instruments, 625 Alpha Drive, Pittsburgh, Pennsylvania 15238; telephone 412-963-2000. Before testing, though, you must drill a ¼-inch hole in the flue pipe (later to be plugged with a sheet-metal screw) to provide access for sampling the flue gases. The absence of such a hole and plug is evidence that the furnace hasn't been tested for combustion efficiency.

If you have a relatively modern oil furnace (one installed later than about 1970), the combustion efficiency should run close to 80 percent. Gas furnaces installed in the last decade or so typically have a combustion efficiency of about 75 percent. Some older furnaces may not be able to reach more than 75 percent (oil) or 70 percent (gas). But if an oil-fired unit has a combustion efficiency of less than 70 percent or a gas furnace less than 65 percent, a tune-up is imperative. An improvement—from 70 to 75 percent, say—would save you $6.70 on every $100 of fuel you buy. As the accompanying table shows, the more room there is for improvement, the more you stand to save.

Increasing Combustion Efficiency

Keep in mind that testing combustion efficiency and improving it are two very different propositions. If, after testing combustion efficiency, you decide that your furnace needs upgrading, call in a qualified service technician to do the work. Tinkering with a furnace is not a do-it-yourself job.

If you have an oil furnace, there are several ways to enhance its combustion efficiency. They include the following steps:

■ Cleaning to remove soot and dirt from the heat exchanger, the flue, and the filters.

■ Sealing cracks and other leaks around the body of the furnace, and tightening loose furnace doors and the like.

■ Adjusting the air intake to be sure the furnace receives the proper amount

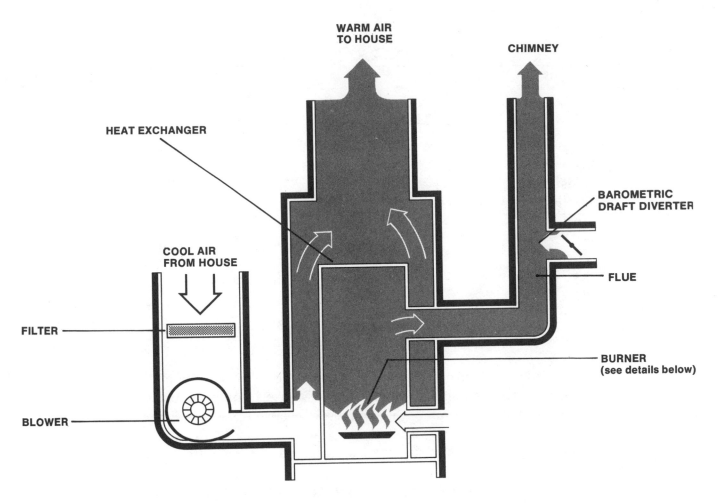

WARM AIR TO HOUSE

CHIMNEY

HEAT EXCHANGER

BAROMETRIC DRAFT DIVERTER

COOL AIR FROM HOUSE

FLUE

FILTER

BURNER
(see details below)

BLOWER

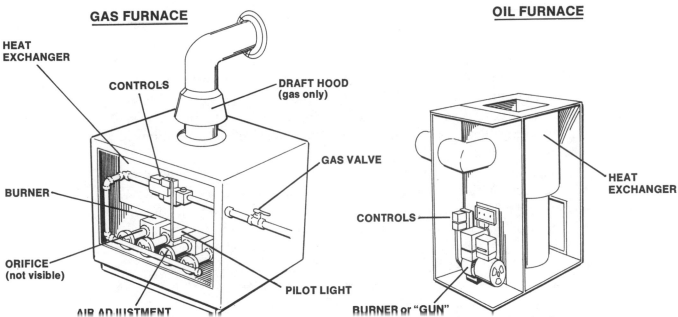

GAS FURNACE

HEAT EXCHANGER

CONTROLS

DRAFT HOOD
(gas only)

GAS VALVE

BURNER

ORIFICE
(not visible)

PILOT LIGHT

AIR ADJUSTMENT

OIL FURNACE

HEAT EXCHANGER

CONTROLS

BURNER or "GUN"

As the large drawing shows, a forced-air furnace draws in cool air from the house, passes the air over a heat exchanger, then circulates it back to the living areas. (Other heating systems work in the same general way, but circulate hot water or steam rather than air.) The smaller drawings point out the important parts of a gas-fired and an oil-fired furnace.

Figure 4.1 How furnaces function

Original Combustion Efficiency	New Combustion Efficiency					
	55%	60%	65%	70%	75%	80%
50%	$9.10	$16.70	$23.10	$28.60	$33.00	$37.50
55		8.30	15.40	21.50	26.70	31.20
60			7.70	14.30	20.00	25.00
65				7.70	13.30	18.80
70					6.70	12.50
75						6.30

SAVING PER $100 OF ANNUAL FUEL COST

of air needed to burn oil completely, plus some excess air to ensure complete combustion. (Too little air will lead to excessive smoke, incomplete combustion, and an inadequate draft up the chimney, whereas too much air will send excessive amounts of heat up the flue.)

■ Installing a nozzle of the appropriate size to provide the optimum fuel delivery rate.

■ Adjusting the oil-pump pressure to meet the furnace manufacturer's specifications.

If you have a conventional gas furnace, there are fewer options available for improving its combustion efficiency. As things now stand, many local building and fuel codes prohibit modifications to a gas furnace or its flue, primarily for safety reasons. Periodic servicing and adjustment and making sure your burner's orifice is clean and properly sized are all useful, but there isn't much more you can do to change a conventional gas furnace's combustion efficiency at this time.

If a gas furnace's efficiency is extremely low, a new furnace may be your only remedy. With oil heat, a very low combustion efficiency may not need so drastic a solution. You may find it feasible to change only part of a low-efficiency oil system—the burner itself.

How can you tell whether your oil furnace is a candidate for a new, high-efficiency burner? You'll have to consider the furnace's flue temperature as well as its combustion efficiency. If the difference between the flue temperature and the temperature of the air in the furnace room is less than 500°F, a device called a *high-efficiency* or *flame-retention burner* could solve a low-combustion-efficiency problem.

If the flue is running more than 500°F hotter than the furnace room, the trouble probably lies in the body of the furnace, rather than in the burner. The furnace's heat exchanger, which transfers the heat of the flame to the air, steam, or water that warms the house, may be too small or too decrepit to do its job well. Baffles inside the furnace, which improve the flame's contact with the heat exchanger, may be damaged or missing.

Sometimes, the flame may be simply too large for the heat exchanger. In that case, derating (by installing a smaller orifice or nozzle) may help improve the combustion efficiency of an oil furnace. Derating is cheap and simple, so it's an approach worth trying. (See discussion of derating later in this chapter.)

Do You Need a New Oil Burner?

You've cleaned your current oil furnace, adjusted the air intake, replaced the nozzle with one of the appropriate size, and adjusted the pump pressure. The furnace's combustion efficiency still hasn't reached 80 percent, but the flue temperature is *less* than 500°F above the furnace room's temperature. In that case, the furnace's burner—the actual firing device—is almost surely the cause of the efficiency problem. A new, high-efficiency oil burner could be a sound investment.

The burner in an oil furnace atomizes oil and mixes it with air for combustion. A conventional burner does a rather poor job of mixing: Some parts of the flame will be fuel-rich, others will be lean. To compensate for this, a conventional burner draws in a relatively large amount of air, so that the fuel-rich parts of the flame burn completely and don't produce an excessive level of smoke.

In an effort to reduce oil consumption and minimize the impact of rising oil prices, a number of high-efficiency burners have been introduced in recent years. These burners (often called flame-retention burners) are more efficient because they mix oil and air more completely. Compared with a conventional burner, a flame-retention model draws in a smaller amount of air and produces a smaller, more compact flame. A flame-retention burner delivers more heat from its flame to the house and reduces the amount of heat that escapes up the chimney.

The flame-retention burner has an added benefit: Because it's designed to draw in only a small amount of air when the furnace is on, the burner also allows only a small amount of already heated air to be drawn up the chimney when the furnace is off. That helps prevent the house from cooling down rapidly, so the furnace needs to run less often. A flue damper (see discussion later in this chapter) does much the same thing, but by actually blocking off the flue when the furnace is off. We estimate that the combination of a flue damper and a flame-retention burner would be half again as effective in reducing this heat loss as either component alone.

A flame-retention burner should present no installation problems that can't be handled by a competent installer. But be sure the installer is familiar with the specific brand of burner you purchase.

It's a good idea to shop for the best price. You may find that some heating-oil companies offer discounts on high-efficiency burners as a way of making oil heat a more attractive proposition.

HIGH-EFFICIENCY FURNACES

Even the most up-to-date heating system will waste some of the energy you buy as heating fuel. But old furnaces—or some new models, for that matter—can be real energy gulpers. Their heat losses put a needless sting in fuel bills.

A modern, high-efficiency-gas furnace (and some oil models, too) can narrow or eliminate one or more of that heat's escape routes. Where an old-style furnace may misdirect as much as half the heat you've paid for, the new designs deliver nearly all the heat content of their fuel to your living spaces. If you now pay $1,300 a year, say, for fuel, a high-efficiency furnace could yield you a saving of more than $400 every heating season. At that rate, even a furnace costing $2,200 or more could pay itself off rather briskly.

Why Furnaces Are Not Perfect

Furnaces—including boilers and other fuel-burning systems—lose energy in several ways while they run; they also lose energy while they are off. On-time and off-time losses are often lumped together when furnace efficiency is discussed, but looking at those losses individually can help clarify the ways that newer designs deliver more heat from fuel.

On-time losses. When a furnace is on, most of its fuel's heat is expended in warming the system's working medium (water, steam, or air), which in turn warms the house. But a part of the heat goes astray. Some goes toward making sure that water created during combustion travels up the chimney as vapor. Some heat other combustion products (chiefly carbon dioxide and nitrogen), which vanish up the flue. And some is used in heating excess air that the furnace takes in to guard against incomplete combustion.

At the same time, the furnace jacket loses heat directly to the surrounding air and to the floor beneath it. A much smaller part of the fuel's energy goes up the chimney as unburned or partially burned fuel—smoke, soot, carbon monoxide, and the like.

Off-time losses. Even after a furnace cycles off, its surfaces continue to dissipate energy to their surroundings and up the flue. A pilot light, if present, also will go on drawing energy and will send it up the chimney in the form of gases and air.

Often more significant—especially during milder months when the furnace is off for relatively long periods—is the escape of previously heated indoor air up the flue. This occurs even when the furnace is cold, as long as the air indoors is warmer than the air outdoors. Per-hour losses from this effect are not high, but a furnace is off for many hours during a heating season.

Measuring Efficiency

To see how these losses affect your pocketbook, you need to know something about a furnace's efficiency. One yardstick is combustion (or steady-state) efficiency (CE or SSE)—the percentage of a fuel's energy converted into usable heat when the furnace runs all the time. As we have already noted, CE is what a technician tests when tuning your furnace. But CE measures only on-time losses, so it doesn't tell you much about a furnace's overall fuel use.

A more realistic measure—and one that the Federal Trade Commission requires for new furnaces and boilers—is a furnace's annual fuel utilization efficiency (AFUE), the percentage of the fuel's energy that can be converted to usable heat over a full year's operation. A furnace's AFUE takes account of both on-time and off-time losses. AFUE figures provide a good basis for comparing furnaces, as well as a rough idea of what you might save in fuel by upgrading your present furnace or buying a high-efficiency one.

If you now own a 20-year-old gas furnace, its AFUE probably is about 55 percent. For a 20-year-old oil furnace, it's likely to be about 60 percent. You can boost these figures a bit by modifying your present furnace. For instance, you can cut off-time losses by having a smaller nozzle installed in an oil burner, so that the furnace runs longer (see the discussion of derating, later in this chapter). We've noted other possible solutions, but they are unlikely to boost the AFUE of a conventional furnace past 70 or 75 percent.

The real gains in fuel-use efficiency have been accomplished by redesigning the furnace itself. In effect, many add-on devices have been designed right into the new units, and the result is a furnace that is capable of consistently delivering an AFUE greater than 90 percent.

Inside a Furnace

When you burn gas or oil, you get carbon monoxide, carbon dioxide, water vapor, and traces of other substances—all of which must be expelled from the furnace. In a regular furnace, those hot by-products are carried up the chimney—along with a lot of heat, contained mainly in the water vapor. Roughly 10 percent of the fuel you burn goes toward keeping that water in vapor form.

A typical high-efficiency furnace is a *condensing* model, which captures much of that fugitive heat with multistage heat exchangers and delivers it to the house. In the process, the furnace's exhaust gases are cooled to 100°F or so, and most of the water vapor in the gases condenses to liquid. That liquid waste, as much as five gallons a day, usually is sent to the sewer or septic system through a drain tube.

Since the other combustion products are not hot enough to get up a chimney on their own, most models expel them with a fan or blower. These models burn their fuel with a continuous flame, as standard furnaces do. However, there are some *pulse-combustion* designs that get rid of their exhaust as part of the combustion process. They burn their fuel in a series of spurts, as an auto's engine does. The pulses force cooled air out through a vent.

As a breed, high-efficiency furnaces are not very quiet. The hot-air version emits both the sound of its circulating fan and either the sound of the fan that provides combustion draft (about the same level as produced by an oil burner) or the hum produced by pulse combustion. A pulsing model isn't notably loud, but a Consumers Union test report characterized its sound as an odd combination of distant bees and a faint hollow resonance. Unless you are particularly sensitive to noise, however, you might not find any of these furnaces especially obtrusive. Nonetheless, you may want to consider acquiring an intake/exhaust muffler kit, which is available for some units.

A modern gas furnace is designed without the energy-wasteful standing pilot light common in older models. Instead, most use an automatic igniter that lights a pilot when needed. Some igniters work a bit like a spark plug in a car.

Optional water heaters are available for certain furnaces. These heaters can take the form of cylindrical, domed tanks about 2 feet in diameter and 4½ feet high. Whenever their stored water cools below a present level, the furnace fires and pumps heated fluid through a coil in the tank to bring the water back up to temperature. This has several advantages: You heat the water at the 90-percent-plus efficiency of the furnace (instead of at the 75-percent efficiency of a regular gas-fired water heater), and the water loses relatively little heat during storage because of the tank's superior insulation.

Let's assume that you install one of these tanks in an area maintained at 70°F, supply it with water at 69°, and fuel it with gas at the 1990 national average price, 56 cents per therm. Heating 600 gallons of water a week to 140°F would then cost you $160 a year. Under similar conditions, a standard gas water heater would cost $235 a year, while an energy-saving standard model would cost $210. What's more, an optional auxiliary tank for a high-efficiency furnace, which quickly replaces the hot water you use, can make full use of the furnace's output—about double that of a conventional water heater.

These niceties, however, come at a price. The hot-water accessory tanks cost two or three times as much as a conventional water heater.

Warranties

The liquid condensate that most of these furnaces produce contains no more acid than a cola drink (pH of about 4), but that's enough to damage metal surfaces and masonry chimneys. In Consumers Union's tests, the gas furnaces, when burning continuously, produced almost three quarts per hour of condensate. One model, an oil type, produced smaller quantities (less than a quart an hour), but the condensate was considerably more acidic—with a measured pH of 2.4, about the level of vinegar or lemon juice.

To avoid acid damage to chimneys, designers often employ 1½- or 2-inch plastic piping to vent the exhaust gases. While you can't safely vent these furnaces directly into a masonry or steel chimney, you can easily route the plastic pipe up through an existing chimney. (Be sure, however, that your local building authorities permit such use.) Better yet, if you are installing a furnace in a new house or relocating the heating plant in an existing one, you need no chimney at all; if local building codes allow, the flue can be run out through the most convenient wall.

The condensate shouldn't harm most sewage systems, including septic fields. And, in most cases, it won't present unusual disposal problems. If you have cast-iron sewer pipes, you may have to add an acid neutralizer to your system. (The neutralizer is a small, inexpensive cartridge that mounts in the furnace's drain line.) If your furnace is located below grade, you might need a small pump to get rid of the condensate. But check with your building officials for local requirements.

Do be aware, however, that acid condensate may attack the furnace's heat exchanger. Furnace manufacturers try to head off problems by using corrosion-resistant alloys and coatings. And well they might—in early models of condensing furnaces, the heat exchangers often rotted out after only a few years of use. Tests sponsored by the industry and the federal government sug-

71

gested an unexpected culprit—contamination by halogens (chlorine and fluorine).

These substances are released by home laundry bleach, other cleaning agents, and aerosol products; once in the air, they eventually can form part of the combustion air drawn in by the furnace. A question naturally arises: What effect might using a washing machine (especially one in a basement laundry room) have on these furnaces? The current crop of furnaces may well have licked this problem.

Consumers Union studied some representative 1987 furnaces to obtain an inkling of how long a new furnace might last and what protection a consumer might have in case of a problem. In too many cases, the warranty language provided cold comfort: The phrasing specifically excluded the most likely cause of failure of the heat exchanger. One warranty, for instance, did not apply if the furnace had "been operated in atmospheres contaminated by compounds of chlorine, fluorine, or other damaging chemicals." A similar proviso lurked in the warranties for other furnaces. Nowhere was it stated whether such unspecified concentrations might include chemicals released by home washing machines and the like.

Although a number of furnaces in the study provided for taking combustion air directly from the outdoors, chlorine traces can also be found outdoors—supplied by nearby dry cleaners, chemical plants, or even the vent of a home clothes dryer. The lesson is: Be sure that the warranty for any furnace you buy will be honored without resort to a "home laundry room" exclusion.

Recommendations

Unless you live in a particularly cold area or have unusually high energy costs, it probably doesn't pay to discard a relatively new, functioning furnace for a high-efficiency model. But if your furnace is more than about 15 years old, it makes sense to install a new, high-efficiency one.

Among high-efficiency models, a comprehensive warranty and a competitive installed price are better buying guides than maximal efficiency. Once you reach an AFUE of more than 90 percent, worrying about slight differences in efficiency becomes almost academic—on a $1,000 fuel bill, the yearly cost difference between furnaces with a 90 and a 95 percent AFUE is $40, at the 1990 national average gas rate. The big saving in fuel comes when you change from a low-efficiency furnace in the first place. The Furnace Size Calculator Worksheet in this chapter will help you determine the potential savings of your installing a high-efficiency furnace.

For most homeowners, efficiency is only part of the picture. Price is just as important, and high efficiency doesn't come cheap. Physical size is much less of a problem—a typical high-efficiency furnace is no larger than an old-fashioned furnace, and much smaller than some.

The unit's heating ability is also a consideration. Luckily, the same design techniques that provide high AFUEs relax the earlier standards of matching these furnaces' heat output to a house's requirements. Assuming that your present furnace's heat input is adequate, start by looking for the rated input, stated on a nameplate on the furnace's cabinet. Then multiply that number by .75 and divide the answer by the AFUE of the furnace you are considering buying. You shouldn't buy a high-efficiency furnace with a smaller input than that—but you can buy a somewhat bigger one, in the event that you might add on to your house.

After you have selected a particular model, the next step is to ask for bids from at least three heating contractors. Be sure to ask for references.

Among the furnaces tested for a January 1987 *Consumer Reports* article, the following gas models had 80 to 84 percent efficiency during the low-demand periods of fall and spring, and 93 to 95 percent when the weather was cold and the furnace would be on almost full-time: Amana HTM Plus EGHW100DB3, Carrier Weathermaker SX 58SX080, Heil Energy Marshal II NUGK075AG, Hydrotherm Hydro-Pulse A100B, Lennox Pulse G14Q480, and Ruud Deluxe 90 Plus UGEA07ECFS. The Amana and Hydrotherm units had optional water heaters available that allowed for heating water at the furnaces' high level of efficiency.

HOW TO CALCULATE FUEL SAVINGS

You can use the worksheet that follows to calculate how much you might save on fuel by installing a high-efficiency gas or oil furnace to replace your present one. A sample set of figures in the left column shows how it's done. (These calculations don't apply if you also plan to switch fuels.)

The example shows a home with an annual gas consumption of 1600 therms. (One therm equals 100,000 Btu, the approximate heat content of 100 cubic feet of gas.) Since that figure includes the fuel to produce the home's hot water, it has been assumed that 80 percent, or 1280 therms, is used for heat.

Another asumption is that the AFUE of the home's present furnace is 55 percent and that the AFUE of the replacement furnace is 90 percent. That means the new furnace would consume 782 therms per year, which represents a reduction of 498 therms. At 56 cents per therm, which was the 1990 national average rate, the annual saving for the home in the example comes to $279.

CUTTING OFF-TIME LOSSES

As noted earlier, combustion efficiency only measures a furnace's ability to extract heat from fuel while the furnace is running. *Seasonal efficiency* measures the percentage of the total energy in the fuel available to heat the house over the entire heating season. While combustion efficiency deals only with on-time losses, seasonal efficiency also takes into account off-time losses.

Some on-time losses are unavoidable. A certain amount of heated air must escape up the chimney to carry away toxic products of combustion. Off-time losses, however, almost never have any beneficial aspects.

Studies have indicated that furnaces typically have seasonal efficiencies of 50 to 60 percent. That is, for every 100,000 Btu of fuel burned by the heating system over the course of a year, only 50,000 to 60,000 Btu ultimately go to heat the house. Warm air continues to escape up the flue even when the furnace is turned off. The annual loss of heat when a furnace is off is roughly equal to the annual loss of heat when the furnace is running.

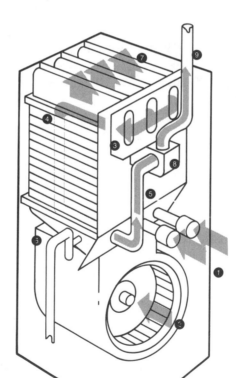

❶ When a high-efficiency hot-air furnace starts up, it takes in air to support combustion at the burners. Some models draw that air directly from outdoors.
❷ When the furnace has heated up sufficiently, a blower starts to turn, taking in cool room air from the house for reheating and circulation to living spaces.
❸ When hot gases leave the burner, they pass through the primary heat exchanger, where they yield some heat to air that will heat the house.
❹ The flue gases, now at about 400°F, enter a secondary heat exchanger, where they surrender still more heat.
❺ By the time the flue gases leave the secondary heat exchanger, they have cooled to about 100°.

❻ As the flue gases are cooled, most of the water vapor they contain condenses to a somewhat acid liquid that is sent through a drain tube to a sewer or septic system.
❼ The heat the flue gases have given up is picked up by the blower-forced air taken in at step 2 and delivered as heating air to room outlets.
❽ Meanwhile, the cool flue gases are forced up a vent by a fan, to be delivered to the outdoors by a chimney or a wall vent.
❾ Plastic vent pipe is used to route the flue gases to their exit point.

Figure 4.2   How a high-efficiency furnace captures extra heat

FUEL SAVINGS WORKSHEET

Amount of Fuel Used Now	Example	Your house
1. Number of units of fuel (therms of gas or gallons of oil) used per year.	1600	_____
2. If some of that fuel is used to produce hot water, put 0.8 here. If none of the fuel is used for hot-water heating, put 1 here.	$\times$ 0.8	_____
3. Multiply line 1 by line 2.	= 1280	_____

Amount of Fuel Used By New Furnace

	Example	Your house
4. Repeat the number on line 3 here.	1280	_____
5. Write the AFUE of your present furnace here. If you don't know what it is, use 55 for gas or 60 for oil.	$\times$ 55	_____
6. Multiply line 4 by line 5.	= 70,400	_____
7. Write the AFUE of the new furnace here.	90	_____
8. Divide line 6 by line 7. The answer is the number of units of fuel you'll have to buy for heating your home with the new furnace for a year.	= 782	_____

Annual Saving in Dollars

	Example	Your house
9. Repeat the number on line 3 here.	1280	_____
10. Repeat the number on line 8 here.	− 782	_____
11. Subtract line 10 from line 9. The answer is the number of units of fuel you can save with the new furnace.	= 498	_____

12. Write the most recent price you
paid for a unit of fuel here.   × $ 0.56 $ _____
13. Multiply line 11 by line 12. The
answer is the number of dollars
you might save per year with
the new furnace.   = $ 279 $ _____

How is this possible? Measured by the hour, off-time losses are rather small. As they accumulate over the course of a heating season, however, off-time losses become generally comparable to on-time losses. That is, for every 100 Btu put in initially, 20 to 25 Btu will be lost while the furnace is firing and another 20 to 25 Btu will be lost when the furnace is off.

The heating capacity of the furnace, compared with the heating needs of the house, largely determines the extent of off-time losses. Furnaces usually are designed to keep a house at some specific temperature (68°F, say) on a *design day*—the coldest day you are likely to have in your locality. (The design-day temperature for Boston is 6°F, for instance, and it's − 16° for Minneapolis.) In the past, however, builders and heating contractors usually installed furnaces with greater capacity than necessary. Fuel was cheap, and a larger furnace was only slightly more expensive than a smaller one. (A larger furnace also would minimize the chances of customer complaints of insufficient heat.)

A furnace that must run continuously to maintain the desired indoor temperature on the design day still will run only 45 percent of the time during an entire heating season. That's because, along with those few very cold days, there are many not-so-cold days. And the warmer the outside air, the less the furnace needs to run. Some excess capacity is desirable to provide sufficient heat on unusually cold days, but bigger isn't always better. The more excess capacity a furnace has, the greater the number and length of its off-time periods and the greater its off-time losses.

In a Consumers Union test, a furnace was set to run continuously. Therefore, there were no off-time losses. Its on-time losses were about 25 Btu for every 100 Btu of fuel. The furnace was then set to run only 15 percent of the time (the equivalent of a furnace too large for its house). On-time losses were still 25 Btu for every 100 Btu of fuel burned, but off-time losses amounted to another 25 Btu per 100.

DERATING

Can anything be done to reduce off-time losses and thus increase seasonal efficiency? The answer is a cautious yes. One realistic approach is *derating*—slowing down the furnace and reducing the number of Btu per hour that it will deliver. By increasing the number of hours the furnace is on, derating decreases the number of hours it is off. While the furnace will run for longer periods, it will use fuel at a slower rate.

Derating can be accomplished by installing a smaller oil nozzle or gas orifice on the furnace. The method works because oil-fired furnaces often have more heating capacity than they need. While many gas-fired furnaces are also larger than necessary, derating usually is not an option for reducing their off-time losses. The reason is generally not technological, but legal: With a conventional gas furnace, derating by itself won't accomplish much toward cutting losses when the furnace is off unless it is accompanied by other modifications, which usually are prohibited under local building and safety codes. Such codes usually require gas-fired furnaces to carry a safety and performance certificate that generally does not cover modifications made after the unit is installed.

To see what effect derating would have on its test furnace, Consumers Union inserted a smaller burner nozzle, reducing the furnace's oil consumption from one gallon per hour of on-time to about two-thirds of a gallon. In order for the furnace to deliver the same amount of usable heat per day that had been the case with the larger nozzle, the furnace ran for longer stretches, but at the lower firing rate. Off-time losses dropped to about 13 Btu for every 100 Btu of fuel. The reduced firing rate, with the resulting reduction in off-time, meant that the furnace needed less fuel to deliver the same amount of usable heat that it had delivered before derating (see the accompanying illustration).

Is Your Furnace a Candidate for Derating?

Even if your oil furnace once was the proper size for your house, it may now have too much heating capacity; insulation, storm windows, and weatherstripping may have reduced the amount of heat your house requires, effectively increasing the furnace's off-time. (Increased use of lights and appliances, which supply some heat, could have further increased the amount of time the furnace remains off.) Consequently, to maximize the saving from insulation and other energy-conserving measures, you should consider whether you need to derate the furnace.

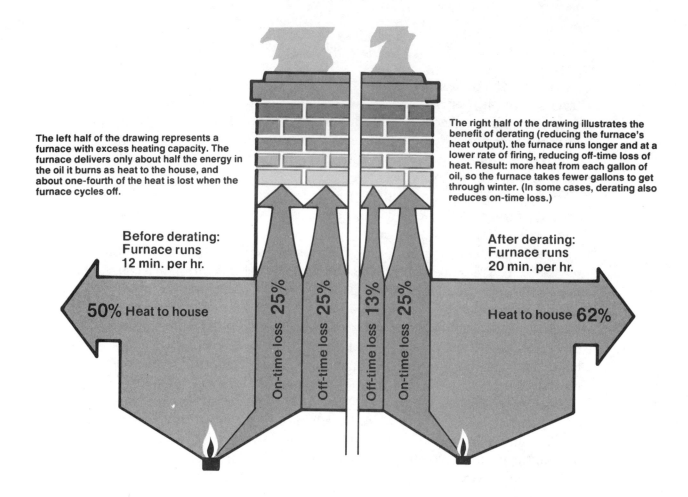

The left half of the drawing represents a furnace with excess heating capacity. The furnace delivers only about half the energy in the oil it burns as heat to the house, and about one-fourth of the heat is lost when the furnace cycles off.

The right half of the drawing illustrates the benefit of derating (reducing the furnace's heat output). the furnace runs longer and at a lower rate of firing, reducing off-time loss of heat. Result: more heat from each gallon of oil, so the furnace takes fewer gallons to get through winter. (In some cases, derating also reduces on-time loss.)

Before derating: Furnace runs 12 min. per hr.

After derating: Furnace runs 20 min. per hr.

50% Heat to house

On-time loss 25%

Off-time loss 25%

Off-time loss 13%

On-time loss 25%

Heat to house 62%

It's difficult to pinpoint how much fuel you could save by reducing off-time losses. The age and size of your oil furnace, the condition of your house, and your family's living habits all affect the amount of on-time and off-time it will accumulate. Certain guidelines will be helpful, however.

For example, Brookhaven National Laboratory in New York found that reducing a furnace's nozzle size by 25 percent could yield a saving of about 110 gallons of oil out of 1,400 gallons burned annually. A different study, conducted in homes in New England, showed that derating always produced a saving—cutting oil consumption by 14 percent, on average, over the course of a heating season.

But not every oil furnace is a candidate for derating. The method is most productive with furnaces that have considerably more heating capacity than they need.

Before you call a service technician to ask for derating, you should calculate whether it is appropriate for your furnace. Use the accompanying worksheet to find out how much excess capacity your furnace has. This will help you decide if derating would be worthwhile.

The method used in this book for estimating furnace capacity is somewhat different from the one used by builders and contractors, but the principle is the same. By measuring your furnace's on-time for a given day, then working through the calculations to account for the differences between that day and a design day, you will be able to derive a size factor for your furnace.

In the example provided, it is assumed that the outdoor temperature is 30°F. The design-day temperature used is 0°F, and the furnace is assumed to fire 12 minutes an hour (20 percent of the time, or 0.2 hour) to maintain an indoor temperature of 68°F. The furnace in the example has a size factor of 2.78—making it a definite candidate for derating. But if the same furnace ran 40 percent of the time (0.4 hour), the size factor would be only 1.40, which is close enough to an optimum size to rule out derating.

Note, too, that derating is not without potential disadvantages. It might increase the wear and tear on a furnace's moving parts, thereby shortening the unit's useful life. It also might entail more frequent replacement of the oil-line filter, or create the need for a second filter. If derating is overdone—leaving the furnace with insufficient heating capacity—it could lead to some chilly days indoors. Derating also could limit the furnace's ability to heat the house quickly in the morning, when indoor temperatures usually are lowest. And derating a furnace that supplies both heat and hot water could leave you with insufficient hot water for bathing and laundering.

Recommendations

The more excess capacity your oil furnace has, the more fuel you stand to save by having the furnace derated.

The actual process of changing the burner nozzle is not difficult. In fact, an oil burner's nozzle should be changed annually as part of routine maintenance. But the job is best left to a qualified technician. To complete the task properly, the technician should check the furnace's combustion efficiency before and after the nozzle is changed, to make sure the derating has not caused an efficiency reduction.

What size should the new nozzle be? As a very rough guide, if the furnace's size factor is about 2, a nozzle 25 percent smaller should work; if the factor is about 2.5, the nozzle can be up to 40 percent smaller; and if the factor is near 3, the nozzle can be as much as 50 percent smaller.

Note, however, that a service technician must take into account other considerations when choosing a new nozzle, including the minimum size specified by the manufacturer. If you have a service contract with your oil dealer, switching nozzles probably won't involve much (if any) additional expense; otherwise, you should expect to pay several dollars for a new nozzle, plus the dealer's hourly labor charge.

CALCULATING WHETHER YOUR FURNACE IS A CANDIDATE FOR DERATING

Use the Furnace Size Calculator Worksheet below to decide if your oil furnace has enough extra capacity so you can install a smaller burner nozzle and use less fuel. You'll have to make a few simple measurements; do them on a cold day when wind conditions are typical for your area. The example we provide will help you follow the calculations.

   1. Measure the temperature inside the house (it's 68°F in our example). Enter the indoor temperature in column 1.

   2. Select a design-day temperature—the coldest day you're likely to encounter in December, January, or February. A design-day temperature of 0°F would be reasonable for most areas north of the Sun Belt; if you live in northern New England, the upper Midwest, or other areas with severe winter weather, use −15°F or −20°F. (We use 0°F in our example.) Enter the design-day temperature in column 2.

   3. Subtract column 2 from column 1. Enter the result in column 3.

   4. Subtract the outdoor temperature on the day you do these calculations from the actual indoor temperature. (In our example, it's 68°F indoors and 30° outdoors, a difference of 38°.) Enter the temperature difference in column 4.

FURNACE SIZE CALCULATOR WORKSHEET

1 Indoor Temperature		2 Design-day Temperature		3 Design Difference		4 Actual Temperature Difference		5 Temperature Correction Factor		6 Furnace on-time		7 Design on-time	8 Furnace Size Factor (1 ÷ col. 7)
68°	−	0°	=	68°	÷	38°	=	1.79	×	0.2	=	.36	2.78
	−		=		÷		=		×		=		

5. Divide column 3 by column 4 to derive a correction factor that will account for the difference between the actual temperature and the design-day temperature. Enter the temperature correction factor in column 5.

6. Measure the amount of time your furnace is on over the course of at least an hour. (If you can measure on-time over several hours, your calculations will be more reliable.) Divide the minutes of on-time by the total number of minutes you monitored the furnace. (In our example, we assume the furnace ran for 12 minutes in an hour; 12 minutes divided by 60 minutes equals 0.2 hour). Enter the furnace on-time in column 6.

7. Multiply column 5 by column 6. Enter the result in column 7.

8. Divide the number 1 by column 7 (in our example: $1 \div 0.36$). Enter the result in column 8. The answer will be the size factor for your furnace.

A size factor of less than 1 means your furnace probably doesn't have the capacity it needs to heat your house on an extremely cold day; consult your oil dealer to see what can be done to improve the unit's performance.

A size factor of 1 to 1.5 indicates that your furnace is the proper size for your house. Derating is not advisable.

A factor of more than 1.5 means your furnace has more heating capacity than it needs. The greater the factor, the more oil—and money—you stand to save by derating.

FLUE DAMPERS

Another way of reducing a furnace's off-time losses—again, usually more possible with an oil-fired furnace than with a gas one—is by having a flue damper installed. The damper, a circular metal plate fitted inside the flue pipe, closes the flue when the furnace is off. Typically, it's operated by a motor linked to the furnace's controls so it pivots open when the furnace starts to fire. Some dampers for gas furnaces have no electrical connections at all. They operate thermally: When the furnace fires and the flue temperature rises, the damper opens; when the furnace shuts down and the flue temperature cools down, the damper closes.

The damper plate doesn't act as an airtight seal. In the case of gas-fired equipment, its looseness provides necessary venting for a pilot light. But no matter how the furnace is fired, the flue should remain open for the early part of every off-cycle in order to clear out the remnants of combustion and to allow the furnace to cool somewhat, once it has stopped firing. If an oil-fired furnace is too hot after firing stops, it can cause oil to dribble from the burner nozzle, which could create excessive levels of smoke and odor.

It's not possible to predict accurately how much fuel a flue damper will save, because several factors affect the off-time losses of any particular furnace. Among them are the furnace's heating capacity, condition, and location; the fuel (oil or gas); the thermostat setting; and the amount of insulation and the number of storm windows in your house.

A number of studies have indicated that a flue damper ought to cut fuel bills between 5 and 20 percent a year. If you have a forced-air furnace that is close to the proper size for your house, your saving will be near the bottom end of that range. (The light construction of such a furnace can't store much heat for long, and the blower usually does a good job of salvaging residual heat from the furnace after it stops firing.) But if you have a hot-water or steam system, and if the furnace has a great deal of excess capacity, your saving will be toward the upper end.

With a gas-fired system, the saving may be smaller than with oil equipment, largely because of the way a gas furnace works. Oil furnaces tend to be more heavily built than gas furnaces of the same type. As a result, they hold the heat a damper traps for a considerable time, making it available for distribution through the house. A design feature of a typical gas furnace makes the heat trapped by the damper available only in the space immediately adjacent to the furnace.

A typical forced-air gas furnace has a secondary air intake around the burner—in addition to air inlets on the burners themselves—to provide enough air to guarantee complete combustion. A gas furnace also may have a draft hood—essentially a large, permanent opening in the flue that helps maintain a fairly constant draft up the flue. If the flue is blocked while the furnace is off, warm air will spill out of the draft hood.

Just as with a forced-air oil furnace, when the burner in a forced-air gas furnace shuts off, the blower continues to run, sending heat in the furnace through the house until the furnace cools down. Because a gas furnace is relatively lightweight, it cools down rapidly.

More important, even when a gas furnace cools down, air from the house will continue to escape through the draft hood. If that air has already been warmed (as it would be if the furnace were located in or near the living area), cooler air from outdoors replaces the warm air, causing the house to cool off.

Having a damper that can block, say, 80 percent of the heat going up the flue isn't the same as saving 80 percent on your heating bills. Remember that on a gas furnace, the flue damper prevents warm air from going up the flue and instead diverts the air out of the draft hood. If the furnace is in or near an area you normally heat, then the damper could be beneficial.

If your furnace is in an unheated basement, or in the attic, a flue damper would save next to nothing. True, it would keep the heat from going up the

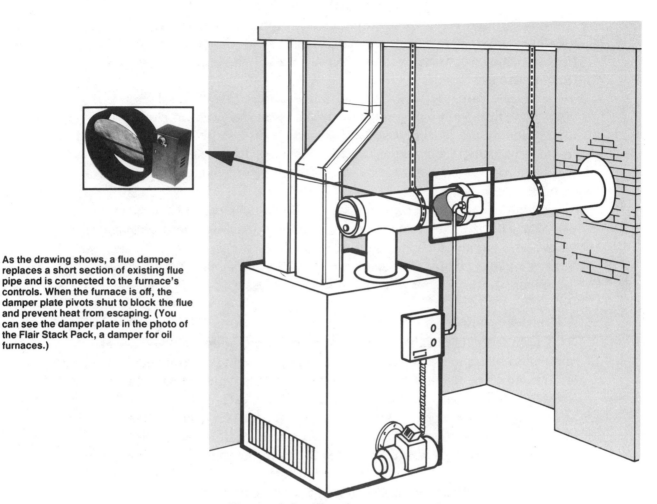

As the drawing shows, a flue damper replaces a short section of existing flue pipe and is connected to the furnace's controls. When the furnace is off, the damper plate pivots shut to block the flue and prevent heat from escaping. (You can see the damper plate in the photo of the Flair Stack Pack, a damper for oil furnaces.)

Figure 4.4   Flue damper on oil furnace

flue, but the damper would divert the heat to an area you don't normally heat, so the reclaimed heat would not do much good.

An oil-to-gas conversion using a power burner eliminates many of the design features that make the use of a flue damper impractical with a conventional gas burner (see "Converting from Oil to Gas," later in this chapter). A power gas burner will behave in much the same way as a high-efficiency oil burner and therefore can benefit more from a flue damper.

Flue dampers are quite safe, provided they are installed properly. Installation, however, may not be a simple matter; installing a flue damper is not a do-it-yourself project. Even if you can buy a unit over the counter, you should be sure to contract for the installation. Check with the installer to see whether the job will affect any warranty on the furnace itself or any service contract you may have with a burner service or utility company. Don't buy a damper unless it carries an Underwriters Laboratories (UL) seal (for oil burners) or an American Gas Association (AGA) Laboratories certification (for gas furnaces).

Work only with a reputable installer. If you are satisfied with the service you receive from your own heating-oil dealer, burner service company, or gas utility, continue with the same one. Or use an installer recommended by friends or neighbors. Before the damper is installed, have the furnace cleaned and serviced so that its controls are in good working order and efficiency is at its peak.

The damper should be mounted solidly in the flue pipe. If the damper is placed in a horizontal section of pipe, it should be attached with sheet-metal screws, and the flue sections themselves should be supported with brackets hung from the ceiling. The damper's control box should be accessible, mounted so heat and soot will not damage it.

Recommendations

As noted, because of the way a gas furnace works, adding a flue damper often offers no appreciable savings. A flue damper might be useful with a gas furnace, however, if yours is a hot-water or steam system rather than a forced-air system, or if the furnace flue is larger than the typical 5-inch diameter. On a hot-water system, closing a flue would cause the furnace to cool down at a slower rate, so that the furnace might still be warm the next time it went on, and therefore less heat would be needed to reheat the water. And, all other things being equal, an 8-inch-diameter flue represents a heat escape route that's much larger than necessary; it incurs more than three times the heat loss of a 5-inch-diameter flue. In situations with such oversize flues, tests have shown that installing flue dampers can reduce gas bills by 20 to 30 percent a year.

A flue damper might help a gas furnace, then, if you want the furnace to warm an adjoining area—as opposed to an area you normally want to heat—if the gas furnace is not a forced-air unit, or if it has an oversize flue. Unless your gas heating system meets these criteria, you probably will be better off investing in other energy-saving devices, such as storm windows, an automatic-setback thermostat, or insulation.

Adding a flue damper *can* be an effective way to reduce an oil furnace's appetite for fuel. In order to realize the greatest saving, of course, you should also have the furnace cleaned and serviced once a year.

Before you buy a flue damper for your oil furnace, you should consider derating the furnace with a smaller nozzle. Derating might even be done with little or no extra charge as part of an annual cleaning and inspection. A flue damper reduces the off-time loss of heat directly, by blocking the flue to keep heat in the house; derating attacks the same problem. With a smaller nozzle installed in the burner, the furnace will run for longer periods, but it will use fuel at a slower rate. Naturally, this will reduce the furnace's off-time and, thus, its off-time losses, while delivering the same total heat.

Remember that if you derate your furnace, you also reduce a flue damper's potential for saving fuel. That, in turn, means it will take longer for a damper to pay for itself through lower fuel bills.

SHOULD YOU SWITCH FROM OIL TO GAS?

Between 1973 and 1980, high prices for fuel oil and promotion by the gas industry convinced almost a million homeowners to convert their heating systems from oil to gas. According to a spokesperson for the American Gas Association, 1,847,000 conversions took place between 1980 and 1989. During these same periods, many other homeowners—influenced perhaps by uncertainty and the cost of conversion as much as by the fuel-oil industry's vigorous counteradvertising—stayed with oil.

How to Evaluate the Claims

The gas and oil industries' conflicting claims of "higher efficiency" and "lower costs" are perplexing. The industry groups, along with various utility companies and oil dealers, have excelled in artful presentation of selected facts: Each side's advertising seems to make its own fuel the inevitable choice for economy. The subject of whether or not to switch fuels needs some perspective.

When you buy a gallon of heating oil or 100 cubic feet of natural gas, you are paying for the actual heat that each can provide to your house. That gallon of oil typically can provide 110,000 Btu; the gas, about 75,000 Btu. Those Btu estimates take into account the combustion efficiencies with which a well-tuned but not-too-new furnace—one installed later than around 1970—can extract the heat: 80 percent for oil-fired equipment and 75 percent for gas.

As for fuel costs: At 1990 national average rates, the gallon of oil would cost 88 cents, the gas 56 cents. So the cost for, say, one million Btu of oil would be $8, against $7.47 for gas. If your local rates correspond to the national averages, heating with gas would afford a slight saving. Two factors should be balanced against any saving, however. The per-Btu price differential between oil and gas is artificial—and probably ephemeral. And, if you already heat with oil, the change to gas equipment could be very costly.

If expenditures to improve furnace efficiency are to be considered investments, converting from oil to gas should be viewed as speculation. If the spread between oil and gas prices holds up over a long period of time, conversion could pay off handsomely.

Recommendations

Investments in fuel conservation will continue to pay off regardless of the price differences between the fuels. Indeed, as you reduce your energy requirements, the price differential becomes less important. Prudent householders should benefit if they concentrate on cutting their energy losses with insulation, caulking, weatherstripping, furnace derating, space heating, and the other techniques described in this book.

In Consumers Union's view, it makes sense to consider conversion only if your existing oil furnace needs replacement or substantial upgrading. Even so, your estimate of how fuel prices will compare in the years after you convert to gas probably will figure into your final decision.

One consideration that used to weigh against oil-to-gas conversion was eliminated as a result of advances in burner technology. Most pre-1980 conversions burned gas at a significantly lower efficiency than the high-efficiency oil burners that were on the market. Although these conversions had allowed a homeowner to heat with a lower-priced fuel, they generally provided little or no increase in efficiency. But now, high-efficiency conversion gas burners also are available.

If you decide that you want to convert from oil to gas, be prepared to spend a considerable amount of money on equipment. The simplest and cheapest conversion is to replace the burner in your furnace. A more expensive alternative is to replace the entire furnace. According to the director of gas energy systems for a large northeastern utility company, a gas burner that would fit into an existing oil furnace would cost about $1,600; a completely new gas furnace for forced-air heat would run about $3,000; and a new boiler for a hot-water or steam system would cost about $3,500. (All costs are 1990 prices and include installation.) To these costs you might have to add $800 to $1,000 to pipe natural gas into your house.

CONVERTING FROM OIL TO GAS

Two types of high-efficiency burners are available for converting an oil-fired furnace to gas: *atmospheric burners* and *power burners*. The two types are quite different in operation, construction, and performance.

The atmospheric burners look and work much like the gas burners usually found in gas furnaces (or gas ranges, for that matter). And they have the advantages of simplicity, reliability, and low cost. However, like older gas burners, they also deliver relatively low combustion efficiency.

An atmospheric burner relies on the pressure in the gas line to force gas through an orifice. And it relies on atmospheric pressure to force air to mix with the gas before ignition (usually by a pilot light). Because mixing is less thorough than it might be, the burner takes in large quantities of excess air. That excess air ensures that the fuel burns completely, without giving off a high level of toxic carbon monoxide. But the excess air also means that much of the heat produced by the burner goes up the chimney rather than into the house. The burner's combustion efficiency—the amount of useful heat produced from each cubic foot of gas—therefore is lower than it could be.

The newer power burners have a higher combustion efficiency than the older, atmospheric types: Power burners use smaller amounts of air and therefore deliver more of their heat to the house. A power burner achieves a better mixing of air and gas—one without much excess air—by using an electric blower to increase the velocity of the air/gas mixture moving through the system. This is the same approach used by the high-efficiency oil burners discussed earlier. In fact, these power gas burners closely resemble the oil burner (or "gun") they are intended to replace.

Eliminating the excess air has another advantage: A power burner can be derated effectively. That is, the burner can be adjusted so that off-time periods—and off-time losses—are minimized. (As we noted in the section on derating, local code restrictions in your area may make that advantage unavailable to you.) An atmospheric burner can't be derated effectively. Another advantage is that some power burners eliminate the standing pilot light, which also saves energy.

You might expect, then, that power burners would be more efficient than atmospheric burners. And they are, according to Consumers Union tests. When run on a specially instrumented oil furnace, the power burners produced combustion efficiencies ranging roughly from 79 to 83 percent—about the same range measured for high-efficiency oil burners. The atmospheric burners ran at about 74 percent combustion efficiency on the furnace. That's typical of most conventional gas-fired equipment.

As for carbon dioxide, a "perfect" burner would have a level of about 12 percent, a sign of complete combustion with no excess air required. Carbon-dioxide levels with the atmospheric burners ranged from 7 to 8 percent. The power burners did better, averaging 9.5 percent.

Although they are complex, high-efficiency conversion burners are similar in design and construction to oil burners, which by and large are reliable devices. A power burner with a spark ignition requires safety controls different from those needed for a pilot-equipped model, but these controls are comparable to the ones found on oil burners.

If your burner is old and outmoded, a high-efficiency gas burner is a feasible alternative to a replacement oil burner. With either type, combustion efficiency will range from 80 to 85 percent. The choice between the two depends on cost.

A new gas burner will cost somewhat more than an oil-fired one if you already have an adequate gas line into your house. Otherwise, installing a gas conversion will cost two or three times as much as installing a replacement oil unit.

If you decide to convert to gas with a high-efficiency conversion burner, find out whether your local utility company will inspect the completed installation.

Consumers Union tested high-efficiency oil-to-gas conversion burners for an October 1981 report. At that time, Beckett, Midco, and Wayne power burners had combustion efficiencies ranging between 79 and 83 percent. Products of those manufacturers are still available, according to the July 1, 1990, edition of the American Gas Association Laboratories list of certified appliances and accessories (AGA Laboratories, 8501 East Pleasant Valley Road, Cleveland, Ohio 44131, tel. 216-524-4990; and 1415 Grande Vista Avenue, Los Angeles, California 90023, tel. 213-261-8161). Of the two atmospheric burners tested for that report, a Wayne unit had a combustion efficiency of 74 percent.

HEAT PUMPS

Regardless of its efficiency, every oil- or gas-fired heating system loses some of the heat from the burning fuel straight up the chimney. Electric heat doesn't squander fuel that way. It's 100 percent efficient. But electricity usually is an expensive source of heat.

Because of the inefficiencies of oil and gas and the expense of electricity, some homeowners may be attracted to the advertised advantages of a fourth type of central heating system—the heat pump, which doesn't burn fuel to produce heat. Instead, it uses the same operating principle as a refrigerator or an air conditioner to extract heat from the air or from the ground and deliver it to the house. Even when the air outside is cold, it contains some heat that can be collected and used to warm the house. Likewise, the earth contains a useful amount of heat, even in winter.

The heat pump's potential for reducing heating bills sometimes can justify the major capital expense required to replace a furnace. But the decision to install a heat pump is more complicated than the decision to replace an aged furnace with a new one.

Air-Sourced Heat Pumps

Until several years ago, nearly all heat pumps were the air-sourced type, making use of the heat in outdoor air. This type often is nicknamed an air conditioner in reverse. In fact, as page 93 explains, many air-sourced models can provide heat in the winter and air conditioning in the summer.

An air-sourced heat pump can operate very economically in areas where the temperature rarely dips below freezing. As the outdoor air temperature drops, however, so does the heat pump's ability to deliver heat. The colder the air, the longer and harder the heat pump must work to extract a useful amount of heat for the house. You would feel that the most when the outdoor temperature dropped sharply or when you tried to warm up an unheated house. For that reason, an air-sourced heat pump is not the type to choose if you live in a colder part of the country.

Designers usually try to compensate for a heat pump's cold-weather deficiencies by building electric heating strips into the system. When a thermostat senses that the unit's normal output isn't enough, the heaters go on automatically. That approach effectively converts the heat pump into an ordinary but expensive electric heater for part of the day.

The need for occasional backup heating also makes it uneconomical to turn down the thermostat on the air-sourced heat pump at night, a common energy-saving strategy with other heating systems. Except in very mild weather, the additional cost of backup electric heat in the morning almost always will wipe out any saving gained overnight.

Ice is another cold-weather problem for an air-sourced heat pump. In near-freezing weather, moisture from the outdoor air condenses and freezes on the pump's heat exchanger, insulating it and worsening the pump's performance. Some heat pumps have electric heaters for deicing. Others run in reverse for a short time to let the warm refrigerant melt the ice. Either method entails an energy cost—electricity to run the heaters or loss of heat from the house.

Water-Sourced Heat Pumps

These models cost a great deal more to buy and install than air-sourced ones, but they can be used successfully in colder parts of the country.

In principle, water-sourcing works the same as air-sourcing, although there are definite advantages to drawing heat from water instead of the air. For one, a given volume of water contains much more heat than the same volume of air at the same temperature. For another, the water is easier to use. All that's needed is a source of water that stays above freezing. (Below the frost line, which is 2 to 4 feet below the surface of the earth, most water stays at about 55°F year round.)

An underground stream or a well does fine as a water supply if:

■ There's enough water. A heat pump may require some 10 to 20 times the amount of water of all other household demands combined. In the New York City area, for example, water consumption would be 10 to 20 gallons per minute.

■ The water is noncorrosive. Acids in some water supplies can damage the heat pump's heat exchanger or other plumbing. Minerals can cause scaling.

■ You can get rid of the waste water. Once the water goes through the heat pump, it must be disposed of. Some local codes may require a discharge well.

Some heat pumps recirculate water (sometimes laced with antifreeze), so they avoid problems of capacity, corrosion, and disposal. These systems, however, require some means of rewarming the water once it has given up its heat to the house. Usually, heat from the earth itself is used. Below the frost line, the earth's temperature is surprisingly stable. Earth is a fairly good conductor of heat, so the water needs only to pass through a sufficient length of piping buried in a trench about 4 feet deep.

"Sufficient length," however, can mean a lot of pipe and a lot of digging. Recirculating installations may involve runs of up to 2,500 feet of 1½-inch plastic pipe. A layout of that length is feasible only on fairly uncrowded, uncluttered land.

Fitting a water-sourced heat pump into a rectangular plot smaller than 100 by 50 feet takes some ingenuity. Two parallel loops, stacked one above the other in a deep trench, can be used. Another alternative is the *geothermal well*—one or more deep holes containing vertical loops of pipe. Digging several shallow wells averts the need for one extraordinarily deep hole.

Comparative Performance

To appreciate the possibilities of a heat pump, you need to know something about its competition. As noted earlier, a typical oil or gas furnace loses at least 15 to 25 percent of its fuel's energy as it operates; that is, the furnace is only 75 to 85 percent efficient. Electric heating loses nothing; it's 100 percent efficient in use. But the heat pump can do better than that.

Consider an electric heat pump taking warmth from outdoor air at 47°F (a

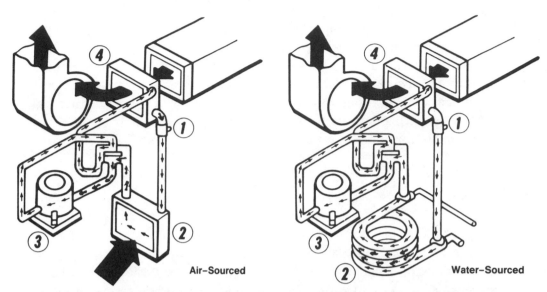

Air–Sourced   Water–Sourced

The key to a heat pump's operation is the refrigerant contained within the unit. The schematic drawings above show how the refrigerant absorbs and then releases heat as it circulates through the system. At the beginning of a cycle ❶, the refrigerant passes through an expansion valve, where it turns from a liquid into a gas, cooling down in the process. In a typical heat pump, the vapor temperature is about 5°F. The refrigerant then flows through a heat exchanger ❷. In an air-sourced unit, the vapor absorbs heat from the outdoor air; in a water-sourced model, it draws heat from a coil of water-filled pipe. The refrigerant is then pumped to a compressor ❸; compressing the gas heats it to as much as 200°. The hot gas then flows through a second heat exchanger ❹, where it gives up much of its heat and turns back into a liquid. Fans and ductwork direct the heat throughout the house.

Figure 4.5   How a heat pump works

standard temperature for rating heat pumps) and using it to heat a house to 70°. Suppose that pump, in continuous operation, delivered 2¾ times the heat of an electric heating system per kilowatt of electricity consumed. (That's a typical level of performance.) In effect, the heat pump would far surpass the electric heater's 100 percent use of its electricity. The heat pump is then said to have a coefficient of performance, or COP, of 2.75. (For all practical purposes, that's the same as saying that the heat pump's efficiency is 275 percent.) An electric heater has a COP of 1.00; a conventional gas or oil furnace's COP might be 0.75 or 0.85.

In actual use, however, COP figures quoted for any heating system can be overly optimistic. The COP ratings used by heat-pump manufacturers assume continuous operation. But heat pumps generally don't run continuously. Every time the unit goes off, hot working fluid in its innards cools, losing some heat to the outdoors. When the heat pump goes on again, it will use some energy to repressurize the working fluid. It must also use energy to deice itself whenever moisture from the air freezes on its components. Finally, the COP ratings used by manufacturers relate to performance only at 47°F. An air-sourced heat pump's COP drops with the outdoor temperature. The heat pump with a COP of 2.75 in 47° weather would have a COP of a bit more than 1.00 if the mercury hit 5°. At that point, the heat pump would be little more than an elaborate electric heater.

A reasonable estimate of a heat pump's performance over an entire heating season is the *seasonal performance factor*. That factor is always less than the theoretical COP, although it can come close in regions where winters are mild. In preparing data for the Heat-Pump Worksheet, the seasonal performance factor for different regions was taken into account.

Costs

Even though a heat pump is a more efficient heater than its competitors, differences in energy prices tend to reduce that advantage. The example used for the worksheet makes the point. The calculations are for a hypothetical homeowner in Bucks County, Pennsylvania, who now heats with oil and pays a theoretical 88 cents a gallon for the fuel. Switching from oil heat to an air-sourced heat pump wouldn't save any money. But our hypothetical homeowner would lose money with a water-sourced heat pump.

Recommendations

A heat pump probably makes the most sense for two types of homeowners: those who want to replace electric heating and those who use more air conditioning than heating.

If you live in an area where winter temperatures seldom drop below about 45°F, an air-sourced unit will serve; one that can handle a 2,000-square-foot house might cost $4,000 to install. Air-sourced models are not an option in areas with severe winter weather. You would need a water-sourced unit with its buried piping, at a cost of at least $5,000. In both cases, ductwork would be extra. The accompanying worksheet will help you estimate whether your saving in fuel would justify the expense of a heat-pump system.

It may take some time to get used to a heat pump if you're accustomed to the hot blast of air and the quick heating provided by a forced-air furnace. Heat pumps deliver air that's only 90° to 100°F, and they heat a cold house relatively slowly.

Be aware that heat pumps, especially water-sourced ones, are a relatively new technology. If you decide to have one installed, you'd be wise to buy it as a package from a contractor who will service the whole unit if anything goes wrong.

Heat Pumps as Air Conditioners

As noted, many heat pumps are designed to lead two lives—heating the house in the winter and cooling it in the summer.

A heat pump devoted only to heating has its condenser indoors to deliver heat and its evaporator outdoors. A dual-purpose unit contains valves that allow the condenser and evaporator to switch functions. The heat pump then works like a conventional air conditioner, drawing heat out of the house.

The industry uses two numbers to characterize the efficiency of a dual-purpose heat pump. The wintertime number is the coefficient of performance (COP), explained earlier. The summertime number is the energy-efficiency rating (EER), or seasonal energy-efficiency rating (SEER), the same number used for other air conditioners. The higher the EER, the more efficient the unit. An often-used variation on the EER is the seasonal energy-efficiency rating, or SEER, which takes into account losses of efficiency over the course of a cooling season. Typically, an air-sourced heat pump has an EER that's comparable with the EER for a conventional central air-conditioning unit. But water-sourced units often have a higher EER.

A dual-purpose heat pump can be an excellent choice for homes—particularly new construction—in the South or along parts of the Pacific coast, where winters are mild but summers are brutal.

HEAT-PUMP WORKSHEET

Would a heat pump reduce your heating bills enough to justify its cost? Would you be better off buying a new, high-efficiency furnace instead? The worksheet below, used in conjunction with the table of fuel factors and the climate map, can help you answer those questions. To guide you through the steps, there are calculations for a hypothetical homeowner in Bucks County, Pennsylvania, who now heats with oil and wants to know if the fuel bill could be cut by installing an air-sourced heat pump.

How Much Heat Do You Now Use?

	Example	Your house
1. Current annual heating bill. Determine the amount you paid for heating last year. Enter that amount here. If your heating system provides both heat and hot water, enter 85 percent of your bill.	$1,110	_____

2. **What does the fuel cost?**
   Enter the price you pay per unit of fuel—gallon of oil, therm of gas, or kilowatt-hour of electricity. If, for example, you pay 7.9 cents per kwh for electricity, enter the price as 0.079, not 7.9.

   $0.88 \qquad \rule{2cm}{0.4pt}

3. **How many fuel units do you use?**
   Divide line 1 by line 2. Enter the result here.

   1,261 \qquad \rule{2cm}{0.4pt}

4. **How efficient is the system you have?**
   Go to the table below and find the fuel factor that's appropriate for the heating system you now have. Enter the factor here.

   98 \qquad \rule{2cm}{0.4pt}

5. **How much heat do you use annually?**
   Multiply line 3 by line 4. Enter the result here. The number you derive is the number of "heat units" (thousands of British thermal units) used to heat the house each year.

   123,578 \qquad \rule{2cm}{0.4pt}

What Would a New Heating System Cost to Run?

6. **Find the fuel factor.**
   From the table above, find the fuel factor for the new type of heating system that you're considering. Enter the appropriate figure here. (The factor is 6.8 for the hypothetical homeowner, who lives in zone 4 and is considering an air-sourced heat pump.)

   6.8 \qquad \rule{2cm}{0.4pt}

7. **How many heat units will the new system use?**
   Divide line 5 by line 6. Enter the result here.

   18,173 \qquad \rule{2cm}{0.4pt}

8. **What will the fuel cost?**
   Enter the price you will pay for fuel for the new system: the price per gallon of oil, or therm of natural gas, or kwh of electricity.

   $0.079 \qquad \rule{2cm}{0.4pt}

9. **What will the system cost to operate?**
   Multiply line 7 by line 8. Enter the result here.

   1,436 \qquad \rule{2cm}{0.4pt}

The amount on line 9 is the estimated annual heating bill for the new type of heating system you're considering. By comparing line 9 with line 1, you can determine how much you can expect to save.

Fuel Factors

Natural gas	60
High-efficiency (condensing) gas furnace	90
Conventional LP gas burner (if LP purchased by the gallon)	55
Heating oil	98
Electric resistance heating	3.4
Water-sourced heat pump	9.4
Air-sourced heat pump (for zone 1 on map)	8.0
Air-sourced heat pump (for zone 2)	7.9
Air-sourced heat pump (for zone 3)	7.5
Air-sourced heat pump (for zone 4)	6.8
Air-sourced heat pump (for zone 5)	6.0

FIVE

Supplementary Heating

If you provide an additional source of heat to the room you use the most, it's possible to cut back on the heat you supply to other parts of your home. Among the options to consider, two involve burning wood: fireplace add-ons and wood stoves. You might also consider portable electric heaters.

In colonial America, heat meant a fireplace or a wood stove. Even as late as 1940, wood was the main source of warmth for roughly one-quarter of American homes. By 1970, though, home-heating systems that used inexpensive oil, gas, or electricity had become just about universal. Wood heated only about 2 percent of American households—mainly those of rural families who could gather their own wood supply.

But the era of cheap energy came to an abrupt end with the oil embargo of 1973. Wood stoves emerged as room and, sometimes, home heaters, with sales exceeding a million a year by 1980.

Despite that temporary surge, space heating with wood didn't threaten the dominance of central heating systems—or even threaten to eliminate other options for space heating. The reason is that burning wood simply is not practical for most people: the tenant in an apartment, the farmer with wheat fields rather than forests, or the owner of a sprawling house too big to heat conveniently with wood. In addition, burning wood contributes to air pollution, as we'll discuss later in this chapter.

Even homeowners for whom wood-burning might make sense would still have to weigh its inconveniences: Part of the backyard must be given over to a woodpile; a stove (or fireplace) will preempt a good part of a room; logs must be hauled home, split, and carried indoors; ashes must be collected and disposed of; the chimney must be cleaned; and the stove or fireplace must be stoked frequently for effective continuous heat production. To heat with oil, gas, or electricity, all you need is a simple adjustment of the thermostat. But with wood, gaining more heat means adding another log, often in the middle of the night.

Such inconveniences don't faze homeowners who hope to slash their home heating costs by burning wood. For them, the question is only which wood-burning route to follow. Should they use a fireplace already in the house? Invest in something to improve fireplace performance? Buy a wood stove—and, if so, which type? Behind these questions lie others: How effectively do the various wood-burning units provide heat? How safe are they? Will burning wood really save money?

For the most part, modern wood stoves and modified fireplaces fill a somewhat different role in home heating than in colonial times. Today these units generally are used in conjunction with, rather than instead of, a central heating system. In effect, many wood-burners are used in much the same way as the electric space heaters described at the end of this chapter. All can make a specific area of your home comfortably warm, permitting you to lower the temperature in other rooms. All save energy and money—but to varying degrees.

FACTS ABOUT FIREPLACES

If you have a fireplace and think you can save energy by building a fire in it, take note: An open fire can draw more heat from the rest of the house than it supplies to the room it is in. Even in that room, the fire's heat contribution may be rather minor; about 90 percent of the heat from wood burned in an open fireplace escapes up the chimney. For every $150 cord of wood you buy, only $15 worth will go toward warming the room.

This is because a working draft comes with every working fireplace. The rush of heated air up the chimney pulls in air from the room around the fireplace—much more air than is needed to keep a flame alive. Strictly speaking, a fireplace doesn't heat the air in a room directly, as a central heating system does. Rather, the open fireplace heats by direct radiation that warms whatever is in the hearth's line of "sight." The closer you are to the fireplace, the more heat you get.

A fireplace, then, directly warms only the side you present to the fire. While one side of your body is receiving heat radiated from the fire, your other side is losing body heat to the room. In colonial times, keeping comfortable often depended on sticking close to the fire and turning oneself around from time to time like a goose on a spit. Eventually, though, the fire would warm the walls facing the fire, which would in turn heat the air in the room. In a modern house, you usually rely on the central heating system to keep the rooms warm. Unlike a fireplace, that system works primarily by heating the air, which in turn heats you and the walls.

A working fireplace in a house with central heating creates an economic problem. The heat radiated by the fire may not be enough to offset the loss up the chimney of room air, which has been preheated by your furnace at considerable expense. Turn off the furnace and the fireplace will indeed heat its immediate surroundings, but the rest of the house is apt to freeze. And most of the heat will go up the chimney in any case.

Fireplace Doors as Energy-Savers

A set of glass doors across the front of the fireplace probably is the best single fireplace device for saving energy. (Doors of almost any material could be used, but glass doors let you see the fire.) Glass doors aren't the most effective heat-savers while a fire burns, but they do cut the loss of centrally heated air after a fire starts to die. At that point, you can't close a fireplace damper lest your house fill with smoke and noxious fumes. (If a fire dies down at your bedtime, the damper might have to stay open all night.) But if you close the glass doors, you will reduce substantially the loss of previously heated air up the chimney.

Most add-on devices for fireplaces include glass doors—or are meant to be used with them. Blowers, vents, and other hardware in these devices are likely to improve a fireplace's performance when compared with glass doors alone.

Glass doors alone are a mixed blessing, however. While they stop the chimney from exhaling too much heated air, they also sharply reduce the heat radiated into the room. The effect can be rather like watching a Yule log on a television set. Still, you can always open the doors when you want to enjoy your fire's warming rays, and shut them when the fire dies down.

When shopping for fireplace doors, look for models with a rigid frame and a seal that goes against the outside rim of the fireplace opening (the thicker and wider the seal, the better). The doors should seal tightly, and inlet dampers should be adjustable. Make sure to get doors that are the right size for your fireplace.

In general, the quality of a set of glass doors corresponds to the price. So does the presence of a few amenities: thermostatic dampers that close themselves when the fire dies down and built-in mesh screens that allow safe use of the fireplace with the doors open.

Recommendations

Most fireplaces—even those equipped with devices intended to save heat—are at best an aesthetic pleasure rather than a serious medium for home heating. If you want to use wood to try to cut home-heating costs, you probably would be better off with a wood stove. But no matter how you use wood for heating, wood-burning has its hazards and inconveniences.

Still, people do use fireplaces, whether for atmosphere or for auxiliary heat. After safety, a fireplace user's first concern should be to reduce the loss of already heated room air up the chimney after the fire dies down. A set of tightly fitting glass doors should do the job. Then, if you expect to use your fireplace primarily for the pleasure of watching a fire, you need buy no more.

If you want to use the fireplace for auxiliary heat, however, you need more than plain glass doors. Before you invest in a fancier device, however, read the sections on wood stoves and electric heaters later in this chapter. Keep in mind, too, that fireplaces with glass doors present some of the same wood-stove hazards as those noted later in the chapter in the section entitled "Safety and Smoke."

There is a way to have a fireplace and also conserve energy. You combine glass doors with a duct to bring outside air directly into the fireplace. Direct venting would virtually eliminate the loss of already-heated room air. Based on a Consumers Union analysis of test data, it was estimated that in a centrally-heated home, direct venting plus closed glass doors could raise the net heat output of a fireplace from about 2000 Btu per hour to about 3000 Btu per hour. Whether it's worth trying to get that type of improvement depends on the cost and complication of rigging a duct from the fireplace to the outdoors; it can be a major undertaking.

If you've been using a grate in your fireplace, there is a cheap and easy way to make an open fire more efficient: Take out the grate. That will produce a slower-burning fire and reduce the amount of air drawn out of the room. An open fire built directly on the fireplace floor in a centrally heated house may come very close to a break-even point in gains and losses, as long as the fire burns briskly. Once the fire dies down, close the fireplace doors to reduce heat loss.

If you're using your fireplace to supplement central heating, turn down your thermostat as far as possible to minimize heat loss. If you want your fireplace to heat a log cabin, don't buy anything but a set of good glass doors. No matter where you live, review the information that follows on wood-burning.

WOOD STOVES

Many old-time stoves were made of cast-iron sections rather loosely joined together with screws or bolts. Such stoves leaked air through virtually every seam. Models made like that are still marketed these days; some may send as much as 80 percent of their heat up the chimney. A better type of stove is "airtight" (which, of course, does not mean it burns wood without air). An airtight stove is one with a sufficiently tight construction so that virtually all air admitted enters through controllable air inlets. It can be adjusted so that a load of wood will burn longer, which will cut your costs for wood. Most of the stoves Consumers Union tested had a tightly fitting door and an air inlet that could throttle down the fire: These are the hallmarks of an airtight stove.

A wood stove works quite differently from a fireplace. In a masonry fireplace, any warming of the room is achieved largely through radiation, which is not a particularly effective nor comfortable method (the side of you facing the fire fries, while the side away freezes). A stove, on the other hand, transmits its heat by conduction through its surfaces; i.e., by warming the room air directly, which then circulates by natural convection. Furthermore, in an open fireplace, you can't control the amount of air that reaches the blaze and steals heat from the flames. About the only control over heat output is in the number of logs you burn. The arrangement makes the open fireplace highly inefficient as a heat provider. In a stove, control of the air supply regulates the rate of burning, which is what makes the stove a relatively efficient way of delivering heat.

In a typical wood stove (see Figure 5.1), room air enters at the front through an *air inlet* that can be opened or closed to control the rate at which the fire burns. Because stoves perform better if fed warm air, the air may pass through a separate *preheat chamber*. From there, the air enters the closed metal *firebox,* where it meets the wood.

In the firebox, the burning wood combines with air to give off heat and flue gases that contain nitrogen, water vapor, carbon monoxide, carbon dioxide, gaseous hydrocarbons, and other waste materials. The whole process is called *primary combustion.*

After leaving the wood, the combustion gases are not always allowed to take the shortest path to the chimney. *Internal baffles* may force the gases to take a longer path and thus release more heat. If the combustion gases and additional fresh air meet at a high-enough temperature, another stage of burning—*secondary combustion*—may occur.

Sooner or later, though, the gases pass through the *flue pipe* (sometimes also called the stovepipe or connector) and into the chimney. En route, the gases may have to pass a *flue damper,* a round plate inside the pipe that's useful for fine-tuning the fire or reducing the blaze to a safe level if the stove overheats.

The flue pipe and chimney needed for a wood stove make up an important—and usually expensive—part of the installation. Occasionally, a homeowner may be able to have installers connect a wood stove to an existing chimney. More often, use of an existing chimney could constitute a fire haz-

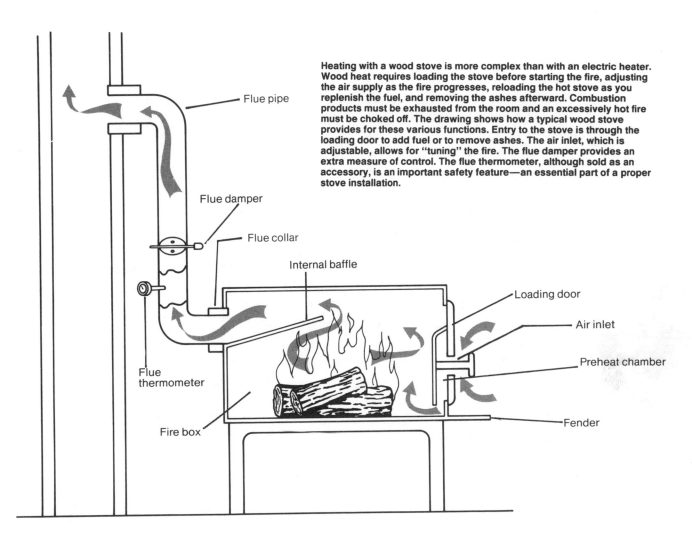

Heating with a wood stove is more complex than with an electric heater. Wood heat requires loading the stove before starting the fire, adjusting the air supply as the fire progresses, reloading the hot stove as you replenish the fuel, and removing the ashes afterward. Combustion products must be exhausted from the room and an excessively hot fire must be choked off. The drawing shows how a typical wood stove provides for these various functions. Entry to the stove is through the loading door to add fuel or to remove ashes. The air inlet, which is adjustable, allows for "tuning" the fire. The flue damper provides an extra measure of control. The flue thermometer, although sold as an accessory, is an important safety feature—an essential part of a proper stove installation.

Figure 5.1   How a typical wood stove works

ard or might violate building codes. It also could mean that the wood stove would have to be located in an inconvenient area of the home. If you are unable to use an existing chimney, installation of a wood stove not only will require a fireproof wall section or insert where the flue pipe will pass through the wall or ceiling, but also will require the construction or purchase of a separate chimney. These possible additional costs could equal the amount you spend on the stove alone. More information about installing a wood stove appears later in this chapter.

Heating with a Wood Stove

As with a fireplace, a stove's heat output depends on how much wood is being burned. But the size and shape of the stove's firebox impose rather stringent limits on the amount of wood you can load into the stove. Some researchers suggest that you merely calculate the volume of the firebox and load the stove to 60 percent of that figure. That's easier said than done, however.

In Consumers Union tests, a number of wood stoves were loaded as full as possible—until the wood pieces reached the top or touched a baffle or other large internal projection. This load, for most stoves, turned out to be only 40 percent of the apparent firebox capacity. That was true even after the wood was split to make the best use of accessible space; gaps between the wood pieces took up the rest of the space.

It's possible, of course, to kindle a fire, adjust the stove so that the flames roar, and stoke the blaze with fast-burning wood every few minutes or so. If you then measure heat output, as some stove makers seem to do, your result will be an impressive figure—produced under unlikely, and perhaps unsafe, conditions.

A more prudent approach is to let a stove's fire build up to a maximum safe level (the maximum output attainable with a flue temperature of 1000°F or less).

The range of heat output from typical wood stoves extends from about 20,000 to 50,000 Btu per hour. (For comparison, a 1,500-watt electric space heater delivers about 5,000 Btu per hour.) For a long, slow fire—the kind you might want while you are asleep or at work—the advantage of "airtight" stoves becomes clear. They have an air inlet that lets you bank the fire properly, thereby permitting control over the rate of fuel consumption.

Safety and Smoke

An overfired wood stove may glow cherry red—an obvious sign of trouble. But even if no external sign of overheating is evident, overheating may be present and damage to the unit may occur. Excessive heat may crack or warp the sidewalls or open joints, or otherwise shorten a stove's life.

As part of its acceptance test program, Underwriters Laboratories checks wood stoves for overheating: A stove is deliberately overfired to see if that will put the stove out of commission or cause nearby combustible surfaces to become overheated.

An overfired stove could also bring on a chimney fire. When wood burns, it gives off creosote, a flammable substance that condenses out of the flue gases as a hard or flaky deposit on the inside of the stove, flue pipe, and chimney. Allowing a stove to burn for weeks with its flue gases at, say, 200°F will promote the formation of creosote. Then, if the flue-gas temprature one day soars to 1,500°F (quite possible when the stove is started or reloaded), a chimney fire is highly likely.

A chimney fire is dangerous. Air rushing through the stove will roar, and the pipe will vibrate and turn red hot. Sparks or flames will erupt from the chimney and imperil the roof. If there are cracks or chinks in the chimney, the flames may well ignite wooden beams or rafters. There is little you can do in the event of a chimney fire but shut down the stove and call the fire department. (Shutting down the stove will cut off only part of the air feeding the chimney fire; leaks in flue-pipe joints will keep it going.)

Improper installation can increase the risk of a fire. (General installation guidelines are discussed later in this chapter.)

A stove should always stand on a noncombustible surface. Use only metal containers for ashes. And be sure ashes are completely cold before mixing them with other trash. Keep an eye on the smoke. A fireplace will spew into the air soot, ash, and some potentially hazardous compounds; wood-burning stoves are even worse offenders. Fireplaces use more air when they burn their wood than do stoves, so the fireplace disposes of noxious by-products more effectively. But adding glass doors or a similar device to a fireplace makes it behave more like a stove.

Smoke from wood stoves contains a good deal of deadly carbon monoxide, organic hydrocarbons, and substances called POMs (polycyclic organic matter)—including some that are known or suspected carcinogens. *New Scientist,* a respected British publication, has characterized the wood stove as "one of the most highly and most dangerously polluting domestic devices known to modern man," but it is still not known whether wood smoke poses a long-term threat to human health. However, wood smoke and haze can be obnoxious enough in the short term.

To reduce smoke and haze, several communities are trying to restrict wood-burning. Vail, Colorado, for example, does not allow more than one wood-fired heater or fireplace to be installed in a new home. The health department of Missoula, Montana, asks residents to refrain from burning wood when air pollution reaches a specified level. Wood is banned as a fuel in a number of cities abroad, including London.

All fossil fuels give off some carbon monoxide when burned. Unlike wood, coal and oil give off sulfur dioxide—a pungent, corrosive gas. Coal also gives off POMs as it burns, and a great deal more ash than wood does.

Here are ways you can minimize the nuisance and potential hazard of smoke from a wood stove.

■ Buy the smallest stove needed for the area you want to heat: If the stove is too big, you'll tend to turn the fire down to a smolder to avoid overheating the room, and the slower the fire, the more smoke it will produce.

■ Use only well-seasoned hardwood, if possible. Green or wet wood burns less completely and gives off more noxious emissions. Hardwoods (ash, birch, maple, and oak, for instance) will provide considerably more heat per cord than softwoods such as cedar, hemlock, and pine.

■ When you start a fire, let the wood burn briskly for 15 minutes or so before you adjust air inlets and dampers to reduce the size of the fire. Don't reduce the air supply to the fire to make the stove burn as long as it can; keep air inlets open at least a crack. If the chimney is still belching clouds of smoke a half hour after you have rekindled a fire, open the stove's air inlet a bit more.

■ Don't overstuff a stove. Big loads of wood create oxygen-starved areas, which tend to burn dirtily.

Air Supply

For any given stove with any given load of wood, a particular amount of incoming air will result in a maximum flue temperature. Any more or less air will drop the temperature of the flue gases by either cooling or starving the fire. The air inlet on a well-designed stove provides that maximum temperature when open about halfway. That arrangement lets you open the air inlet fully to start fires, close it fully to extinguish them, and adjust the stove to accommodate large and small loads of wood.

Unfortunately, you can't always tell whether or not a stove's air system is well designed until the stove is in use.

Because you can't tell how an inlet will work just by looking at it, a *flue thermometer* is an indispensable safety accessory; it should be installed as close as practical to the stove's *flue collar*. A safe temperature range is 350° to 1,000°F for steady operation. Temperatures below that range encourage the buildup of creosote. (The Bacharach 1207018 Tempoint thermometer, about $42.50, should work well. It's available from some heating supply houses and large hardware stores. Check with Bacharach Instruments, 625 Alpha Drive, Pittsburgh, Pennsylvania 15238, tel. 412-963-2000.)

Stove Construction

Fortunately, you can judge for yourself some details of a stove's quality. Here are various factors to consider:

Firebox. In the store, look closely at the firebox: The life of your stove probably depends on it. Bare cast iron and steel are susceptible to rust. Cast-iron stoves—even "airtight" ones—are made with many joints that could develop leaks and would need to be patched with furnace cement.

Consumers Union prefers those made of plate steel more than ⅛ inch thick, with continuously welded joints; such joints are unlikely to separate. Stoves with sheet-metal walls less than ⅛ inch thick are apt to warp after they have been heated a few times. That may not pose a structural problem, but it will affect the look of the stove.

The thinner the metal (whether cast iron or steel), the faster it will have a hole burned in it. So be sure to ask about the thickness of the stove's metal. Tapping a stove can provide a quick check: Thin metal *sounds* tinny.

Interior features also can add durability. Internal heat shields or baffles at the sides can protect stove walls from the searing heat of the fire. A firebrick lining is a good feature.

Fittings. Just opening a stove's door and adjusting its air inlet a few times can give some inkling of quality. Doors should shut and latch firmly, and they should not sag on their hinges once opened. Air inlets should adjust easily but not loosely. Such checks not only yield an idea of workmanship but also indicate the potential for leaks.

Check for the tightness of a door's seal by closing the door on a piece of paper; if you can slide out the paper, the stove probably has a leak. But the only sure test for leaky fittings is a smoke test: If a sample of the stove you want is operating in the dealer's showroom, move a lighted candle or match around the stove and see where the candle's smoke is drawn in. If you're unable to do the test, you can only hope that the stove is actually airtight.

Remember that door gaskets are no assurance of a tight door.

Convenience

A wood stove may require tending at odd hours and a certain readjustment of lifestyle. Still, some stoves make life easier than others. Here are some points to consider:

Loading. Ease of loading should take top priority among convenience factors. Once you've decided on the size of stove you need, compare several competitors for the length of log they'll take, usable internal volume, and sizes and locations of loading doors. These factors, by the way, must be considered together.

Door opening. Handles should be insulated (a wooden knob or a springlike contrivance is typical), or a separate opening tool should be provided. If a stove comes with such a tool, check that it engages the door firmly.

A stove's door must be opened with great care. If you open it quickly, the inrushing air may create an exhalation of smoke, perhaps accompanied by a face-searing flare-up. Before opening a door, open the air inlet, then keep the door open a crack for a few seconds before opening it fully.

Air inlet. Every time you start a fire, you'll have to adjust the air inlet a number of times. Inlet controls often get too hot to handle. If a stove doesn't provide a tool for adjusting the air inlet, you'll have to provide one yourself, or use an oven mitt. Check any tool that's supplied for easy operation and proper fit. While you're at it, check the security of the air inlet itself.

An automatic air inlet is a nicety. The initial setting, done by hand, can be difficult and tedious. But after that, a bimetallic strip takes over to help keep an even temperature in the firebox. (The strip is similar to the coil found in most thermostats. It expands and contracts according to temperature changes. That movement automatically opens and shuts the air inlet.)

Ash removal and cleaning. Even the most devoted stove owner doesn't enjoy cleaning out ashes. The job is easiest with stoves that let ashes fall through a grate into a removable drawer or chamber below the firebox.

Recommendations

First ask yourself if you *need* a wood stove to help lower your home-heating costs. Are you sure you have done as much as you can to save energy in other ways? Will you be able to get the wood you need, when you need it? Do you have a place to store the wood? Are you fully aware of all costs of installation (possibly including a separate chimney)? Are you prepared for the inconveniences, major and minor, that a stove entails? To what extent do you plan to rely on a wood stove for heat? If you're concerned more about function than charm, how about using an electric heater or two instead?

You might consider a stove that's suited for heating only one or two rooms as a supplement to your central heating system—on a par with several electric space heaters, for example. A batch of heaters, of course, would be more convenient to operate and less expensive to buy than a wood stove. The stove, for its part, might well be cheaper to run, depending on the cost of electricity and the cost of wood in your area. Some of the economic factors are noted in greater detail in the section that follows.

As far as heat output is concerned, the type of stove you choose—upright, box, potbelly, or whatever—matters less than the physical size of the stove and the quality of its construction. As you would expect, the larger the stove, the more heat it can produce. Construction details—a tightly fitting door, an air inlet that allows a wide range of adjustment, the absence of obvious cracks around the firebox—all denote a stove that can deliver a good, slow fire instead of sending heat up the chimney.

Give preference to a stove made of heavy plate steel, with a tightly fitting loading door, a removable ash container, and an air inlet that works smoothly. By all means, avoid cheap stoves made of thin metal; these are apt to be unsafe.

Installing and Maintaining a Wood Stove

Poor installation is a frequent culprit in home fires related to wood burning. Installing a wood-burning stove is not a do-it-yourself job; leave it to a professional. But be sure to take this cost into account before deciding on a wood stove in the first place. Installation of the stove and the kind of chimney you might need could more than double the cost of the stove.

Before contacting the professional installer, check with the local building or fire department to find out the requirements in your area. Also contact your insurance company to find out if the presence of a wood stove will affect your insurance rates. Once the installation is complete, ask your insurance company and local fire and building authorities to inspect it to be sure it conforms to safety regulations.

Chimneys must be inspected as well. Ironically, the very inefficiency of an open fireplace has a protective function. The constant flow of air up the chimney slows down the accumulation of combustible creosote on the chimney walls. More efficient devices, such as airtight stoves, can slow down fires significantly, giving creosote an opportunity to build up more quickly. If you burn slow fires as a matter of course, a chimney fire is a serious risk.

Before the installation of a new fireplace add-on or a stove, inspect the chimney. By shining a flashlight up the chimney and holding a mirror at the proper angle, you'll be able to see into the blackness right from the fireplace. In the case of a chimney that has a stove already attached, the chimney may have to be checked from the roof. Call in a chimney sweep if you find tarry or oily creosote deposits that are more than about ¼ inch thick. Once a fireplace device has been installed, the chimney should be inspected regularly (about every month) to be sure the creosote is not building up. With a stove, check every two weeks until cleaning is indicated, and then establish a cleaning schedule.

Cleaning and inspection are also in order if you plan to reactivate an unused fireplace or chimney. Older chimneys may have deteriorated dangerously, or may never have been built properly in the first place. Your local fire or building department may be willing to do the inspection or recommend someone to do it.

If you decide to add fireplace devices, your installation will have to be wherever your fireplace is located; a wood stove, on the other hand, may offer you some options in location. Those options, however, are far from unlimited: A stove placed too close to an unprotected wall creates a fire hazard. The accompanying drawings will assist you in planning a safe installation, but they provide only general guidelines, since local building and fire safety codes vary, and so do individual houses. Your stove installation should be tailored to the construction and layout of the house, not to a generalized set of specifications.

Facts and Figures About Wood-Burning

If you could use a wood stove as the primary source of heat for your home, then figuring the saving from wood heat would be straightforward. On the assumption that a homeowner uses 1,000 gallons of oil to get through a winter, at the 1990 national average price of 88 cents per gallon, the oil bill would have been $880. To get the same amount of heat from wood, the homeowner would need about 10 cords of hardwood burned in a stove that was 50 percent efficient. If wood cost, say, $125 per cord, the homeowner would have paid $1,250. At 1990 average fuel prices, the wood was priced $370 higher than the heating oil. And if wood cost $200 per cord, the homeowner would have paid $1,120 more for wood than for oil.

However, most stoves supply enough heat for only one or two rooms. In effect, a stove should be considered as an auxiliary heater, serving the same function as electric space heaters. A small stove can deliver about the same amount of heat as four 1,500-watt electric heaters set for maximum output (roughly 20,000 Btu per hour).

Using either a stove or an electric heater could lower your overall annual fuel bill. Whether from electricity or wood, the auxiliary heat would allow you to run the central heating system less often, or at a reduced temperature level. But the overall saving isn't predictable: So much depends on fuel prices and, more important, on the amount of central heat and auxiliary heat you require.

Let's suppose that you are trying to choose between a stove and portable electric heaters to heat two rooms. A small stove, run for nine hours a day, would use less than $2.50 worth of wood (assuming that the wood costs $125 per cord). You would need four 1,500-watt portable heaters to get the same amount of heat. In nine hours, the heaters would use about $4.20 worth of electricity (assuming that the electricity cost the 1990 national average of 7.9 cents per kilowatt-hour).

Based on that example, a stove would seem to be a fairly economical supplement to a central heating system. Of course, where wood is cheaper or electricity more expensive, a stove would be a very economical space heater. But there are other factors. A good wood stove is more expensive than 4 por-

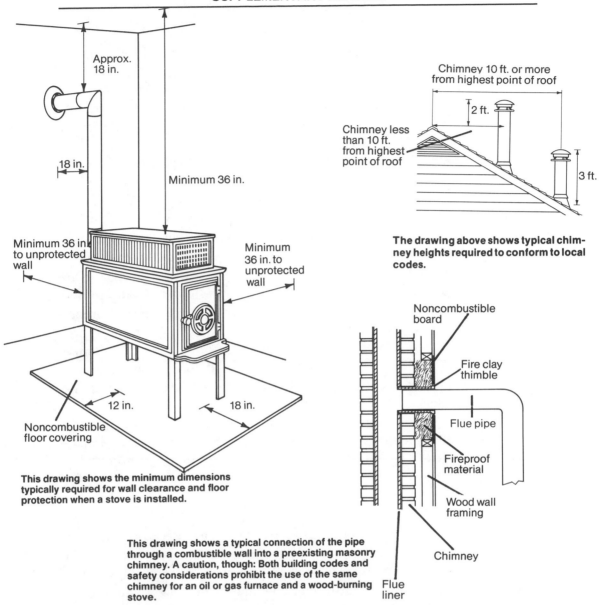

Approx. 18 in.

18 in.

Minimum 36 in.

Minimum 36 in. to unprotected wall

Minimum 36 in. to unprotected wall

12 in.

18 in.

Noncombustible floor covering

This drawing shows the minimum dimensions typically required for wall clearance and floor protection when a stove is installed.

Chimney 10 ft. or more from highest point of roof

2 ft.

Chimney less than 10 ft. from highest point of roof

3 ft.

The drawing above shows typical chimney heights required to conform to local codes.

Noncombustible board

Fire clay thimble

Flue pipe

Fireproof material

Wood wall framing

Chimney

Flue liner

This drawing shows a typical connection of the pipe through a combustible wall into a preexisting masonry chimney. A caution, though: Both building codes and safety considerations prohibit the use of the same chimney for an oil or gas furnace and a wood-burning stove.

Figure 5.2 Safe installation: Keeping the fire in its place

table electric heaters to buy and install, so it would take some time before you could recover your costs through lower fuel bills. A stove also is far less convenient to operate than an electric heater.

Quirks in wood marketing can also complicate your cost calculations. The standard measurement for firewood is the cord—a tightly packed stack of logs measuring 4 by 4 by 8 feet and weighing about 2 tons. Watch out for the "face cord"—a 4-by-8-foot stack that looks like a full cord from the front but is only a foot or two deep. A face cord should be one-fourth to one-half the price of a full cord. Watch out, as well, for wood sold by the "truckload"; until it's stacked, you have little idea how much you are getting. (After all, a half-ton pickup truck can't carry a 2-ton cord of wood.) And be sure to buy only well-seasoned hardwood if you can get it; the logs should have dark ends, with cracks radiating from the center. Unseasoned wood will give you a lot less heat. So will a cord that contains some softwood.

If you're using a fireplace for heat, note that eliminating the grate is not the only way to improve an open fire. Pile the logs at the back of the fireplace, with the largest log at the back. The fire will then warm the back wall, the biggest log will act as a reflector, and the net effect is that extra energy will be radiated toward you. Open the fireplace damper only enough to keep the fire from smoking; leaping flames are a sign of too much air. Build small fires—the flames shouldn't extend up behind the lintel—the structural cross member that supports the structure above the fireplace. If you use an ordinary grate, leave plenty of ashes under it to help cut airflow and to keep the embers warm longer.

To reduce the amount of air the fireplace steals from the rest of the house, close any doors that open onto the fireplace area and open a window a crack near the fireplace. (A window on the same wall as the fireplace is best; one opposite will create a draft that sweeps the room.)

If a wood stove is your choice, be sure not to overload or overfire it. Never let a fire roar until the stove's metal glows; install a flue thermometer and use it to keep the stove's operating temperature between 350° and 1,000°F.

PORTABLE ELECTRIC HEATERS

At average rates, a 1,500-watt portable electric heater uses about 12 cents worth of electricity an hour. In some parts of the country, where electricity is more expensive, the operating cost may run to about 18 cents an hour. That may seem high, but the cost of running a portable heater pales beside the cost of fueling a central heating system—the single biggest user of energy in most homes. A gas or oil furnace can run to well over a dollar an hour.

The opportunity arises for energy saving, at no sacrifice in comfort, when you need to heat only one room or part of a room. At this point, if you use the portable heater and lower the setting on the central system's thermostat, you're apt to save energy—and money—in the long run. And, unlike the central heating system, all electric heaters convert into heat virtually 100 percent of the power they draw—regardless of the size, shape, or type of heating element used.

Thus, you might use a portable heater to warm the kitchen during breakfast while leaving the central system's thermostat at a low nighttime temperature for an extra hour or so. Or, if you've turned off the central heat as soon as possible after winter, you might use a heater to take the chill off an early spring day.

Of course, a wood stove or a fireplace can afford similar opportunities. But wood-burners are expensive to buy, inconvenient to start up, and rooted to the spot by their dependence on a chimney. You can buy a good electric heater for between $50 and $80, depending on the type, set it to work at the flick of a switch, and move it from room to room as needs dictate.

A typical portable heater has a power rating of 1,500 watts. Because that amounts to most of the capacity of an ordinary 15-amp household electric circuit, you can't use any other heating appliance simultaneously on the same circuit. Some heaters are rated at about 1,350 watts, or have more than one heat setting, thereby providing added flexibility for its use. (Of course, at a lower wattage setting, there's less heat.)

The most significant performance aspects among the various types of heaters available are apt to be how well a given room temperature is maintained and how evenly heat is distributed. Electric-element, liquid-filled, and ceramic heaters are primarily convective. They move hot air around a room either with a fan or by encouraging natural convection currents. Two types, radiant heaters and quartz heaters, emit most of their energy in the form of infrared radiation from glowing heating elements.

Heater Types

Convection heaters. The basics of convection heaters have remained unchanged for decades. An internal fan blows room air over heated wire or coil elements. New touches include modernistic styling in a heat-resistant plastic housing, smaller outside dimensions, and the addition of such gadgets as electronic timers. Available in either upright or baseboard formats, convection heaters have the advantages of portability, low price, and mechanical simplicity. They warm up fast, but they usually are not very good at spot heating. Convection heaters become very hot; they should be used cautiously and kept away from all fabrics.

Liquid-filled heaters. An electric element sealed in the unit warms either water or oil, and the hot radiator creates natural convection—circulating air currents—in the room. The thermal mass of the liquid holds heat, which helps to minimize room-temperature fluctuations. These heaters tend to be bulky, heavy, and difficult to move. They are slow to warm up, and the upright type can be toppled easily. Their heat distribution is generally mediocre, and they are not good for spot heating.

Ceramic heaters. This relatively new type of heater also warms by convection. A special electrically conductive ceramic element heats quickly but then levels off as it approaches its maximum operating temperature. Heat is delivered by a fan that draws air through the ceramic. Ceramic heaters are fairly safe because their elements never become hot enough to start a fire. They are compact and lightweight. A drawback is that they are relatively unresponsive to changes in a room's temperature. Either they allow wide temperature swings or they can't keep up with changing outside temperatures.

Radiant heaters. The principle here is that a glowing heating element in front of a shiny reflector radiates heat, warming people and furnishings rather than the intervening air. Most models have a fan, which converts some of the heater's energy to forced convective heat. Radiant heaters are good at spot heating—warming a person quickly even in a cold space. A radiant heater is the right choice for people who work in a large, unheated, or poorly heated area. Some of these heaters pose a fire hazard, however, because they focus so much heat in one direction. Convection heaters are safer in this resepct.

Quartz heaters. These radiant heaters use an electric-coil element enclosed in a quartz glass tube. They may also have a fan. During the night, a quartz heater's bright glow could annoy a light sleeper; it's best at spot heating, but it may pose a significant fire hazard because of the narrowly focused, intense heat it produces.

Finding the Right Model for Your Home

Checks of convenience will disclose a variety of variations among models that may be important to you. Here's what to look for:

Ease of turning on and off. Most heaters are not very tall, so the handiest place for the on/off switch is on top; switches in other locations will be less accessible. The most convenient on/off switch is one that is independent of the thermostat control. That way, when you turn the unit off, you can leave the thermostat at a setting you have found satisfactory. Try to avoid a model that is shut off only by turning the thermostat down; because some of these lack a positive off control, there's a chance that the unit will continue to heat even though you think it's off (unless you unplug it when it's not in use). To find out whether the heater's thermostat has a positive off switch, listen for an audible click as you turn down the dial. If you don't hear such a click, choose another model.

Ease of selecting settings. If a heater gives you a choice of wattage settings, check the control for clear markings. When you operate the heater, use the lowest wattage setting that keeps you comfortably warm.

Quietness. Electric heaters tend to be rather noisy, especially when they are first turned on or when they are cycling. In normal operation, heating elements may buzz, fans may hum, and metal parts may click and pop as they expand (when they heat) and contract (when they cool). As a rule, heaters with fan-forced airflow are apt to be the noisiest. If noise is of particular concern to you, be sure to listen to a heater in operation before you buy it.

Ease of carrying. Size, weight, balance, and overall dimensions are all factors to consider, but the importance you attach to each may well depend on whether you want to move the heater often or use it in only one place.

Nonabrasive base. You may encounter a heater with unshielded metal feet or an otherwise sharp-edged base that could scratch the floor. Check with a bare finger—applied carefully—to find any such problems.

Safety

Some kinds of portable heater can singe a carpet or set it afire if they are tipped over. All manufacturers guard against this by one means or another. Many heaters include a built-in switch that automatically turns off the heater if the unit falls forward or backward. In some, this tip-over switch goes into action if the heater is tilted in any direction. Other heaters safeguard you even further by including overheat-protection sensors that shut off the unit if its internal temperatures become too hot.

Some models have a grille or housing that can become too hot to touch even though it isn't hot enough to start a fire. In others, a potential for electric shock exists if children poke their fingers or a metallic object through the grille. (This problem is reduced in heaters that have a fine-mesh grille.) To rule out any chance of shock from a heater that's not in use, always turn off the heater and unplug it. (Turning off the heater first will prevent sparks when you pull out a plug.)

Despite the inherent potential for a heater to present hazards, most models are safe if they are used properly. But caution is still the watchword. Don't set a heater where it can be knocked over easily, and keep it away from draperies and other fabric-covered objects. Never try to dry clothes by draping them over a heater. If you must use an extension cord, use the heavy-duty type (at least 16-gauge wire), and don't run the cord under a carpet.

Recommendations

Ceramic heaters don't have much to offer besides their compactness. A convection heater is the best choice for warming a room quickly and keeping it warm. Radiant heaters are the best choice if you just want to keep yourself warm—in a chilly workshop or office, for example—and need not be concerned about raising the temperature of the whole space.

In selecting a convection heater, look first to a medium-sized or compact model. The best of them work very well, overall, and are small and light enough to be truly portable.

Large, liquid-filled convection heaters lend themselves to semipermanent installation in a room that needs supplementary heating. Some are low-slung units nearly 6 feet long. Others are the size and shape of an old-fashioned radiator, and, like baseboard heating elements or radiators, these electric heaters typically contain heated oil or water that warms the room air. (For Ratings of electric heaters, see Appendix B.)

SIX

Water Heating

Of all residential appliances, the water heater is among the greediest for energy: Only a central heating or air-conditioning system demands more. According to the U.S. Department of Energy, a water heater consumes some 20 percent of the energy used in the home. Changing energy costs probably haven't altered patterns of hot-water use significantly, making it a virtual certainty that, for most households, heating water is expensive.

The total cost of heating water includes the cost of heating and the cost of storage (or standby) losses. There is also the initial equipment cost. We will consider each of these three major costs separately.

HEATING WATER

It takes energy to raise the temperature of water. The more water you want to heat, and the hotter you want it, the more energy you need. There is a simple formula to express the interrelationship of these factors: 1 Btu will raise the temperature of 1 pound of water by 1°F. To raise the temperature of, say, 15 gallons of water (about 125 pounds) from 60° to 140°F would require 10,000 Btu (125 times 80).

While that Btu requirement remains the same regardless of the source of the energy, the *cost* of the energy can vary considerably. At one extreme, we have the energy cost of using a solar water heater: As long as the sun is shining, the energy cost of heating 15 gallons of water is essentially zero. To heat the same 15 gallons, an electric water heater would require about 3 kwh, or 24 cents, at an average electricity rate of 7.9 cents per kwh. A gas water heater, which loses some of its energy up the flue, would have to burn 13,000 to 14,000 Btu—about 8 cents worth of gas—to deliver the 10,000 Btu required to heat the 15 gallons of water.

No matter what type of water heater you have, there are two simple ways to cut the operating cost: Lower the temperature of the hot water you use, and use less hot water.

If water flows into an electric water heater's tank at about 60°F, heating 64.3 gallons of water* from 60° to 140° would require about 13 kwh, which would cost about $1.04 (with electricity billed at 7.9 cents per kwh). If you lower the heater's thermostat setting to 120°F, however, you reduce the cost of heating that 64.3 gallons by about 25 percent, or 26 cents. (See the accompanying table for information on savings with other temperature settings.) And you reap a safety benefit: You reduce the risk of scalds.

In reality, though, the temperature setback probably wouldn't yield a full 25-percent saving on your water-heating bill. It's usual to draw a mix of hot and cold water for bathing, showering, washing, and the like. Lowering the hot-water temperature would simply mean that you would use more hot water and less cold for your bath by opening your hot-water faucet more and cutting back on the cold. But you would be able to save on the amount of hot water going to a dishwasher or washing machine.

To some extent, turning down a water heater's thermostat is a matter of trial and error. Set the temperature too low and a dishwasher (if it's not the kind with a built-in water heater) might not do a satisfactory job. To find the right setting, start by lowering the thermostat slightly. If you still have enough hot water for your needs, lower the setting further. Each time you lower the thermostat, check the performance of your dishwasher. When you reach the point where the water is no longer hot enough to wash dishes satisfactorily, raise the temperature setting a few degrees. If you don't have a dishwasher, use your judgment about whether the household's hot-water needs are satisfied.

The second—and more obvious—way to reduce the cost of heating water is simply to use less. Effective, if mundane, tactics include: taking shorter showers, using less bath water, substituting warm or cold clothes-washing cycles for hot, and postponing use of a dishwasher until it's full.

Flow restrictors and low-flow shower heads can be very helpful in reducing your use of hot water for showering. In effect, flow restrictors that are inserted in a regular shower head set a limit on the water you obtain through that head. Low-flow shower heads, for their part, let you cut down on the water you use for showering, usually without unpleasantly enfeebling the flow. (See the dicussion of these devices at the end of this chapter.)

STORING WATER

A conventional tank-type (storage) water heater is almost always filled with hot water. These tanks are invariably insulated, but an inexorable loss of heat still occurs. Unlike losses during heating, storage losses are largely unaffected

*The daily use by the average American family, based on the Department of Energy's estimate of 450 gallons used per week.

ANNUAL COST OF HEATING 64.3 GALLONS OF WATER PER DAY (EXCLUDING STORAGE)		
Temperature Raised from 60°F to—	**Gas Water Heater Dollars/Year***	**Electric Water Heater Dollars/Year***
160°F	$143	$449
150°	129	403
140°	115	358
130°	100	313
120°	86	269

*Gas at 56 cents per 100 cubic feet; electricity at 7.9 cents per kwh.

by the amount of water you use. To cut storage losses, you need to reduce the difference in temperatures between the water inside the heater and the air outside, increase the effectiveness of the heater's insulation, or both.

Efforts to narrow the temperature difference between the hot water and the surrounding air usually focus on lowering the temperature of the water in the tank. (If you are installing a new water heater, however, be sure to avoid locating it in a garage or other unheated area.) If the average temperature of the room in which the water heater is located is, say, 70°F year round, then resetting the heater's thermostat to drop the temperature from 140° to 120°F would lower standby losses almost 30 percent. (See the accompanying table for a summary of the savings possible by reducing storage losses.)

How much storage losses will actually contribute to your cost of operating a water heater depends on how well insulated your unit is. A well-insulated electric water heater will lose about $62 worth of heat in a year; a poorly insulated one will lose perhaps twice that. A well-insulated gas model will lose about $32 worth of heat annually, as opposed to $52 with a poorly insulated gas heater. (The reduction in storage losses is less marked with gas models because some of the heat loss is up the flue, and that particular heat loss can't be helped by insulation.)

BUYING EQUIPMENT

As noted, in addition to the costs of heating and storing water, there is another major factor to consider—the initial outlay for the heating and storing equipment and for its installation. With conventional water heaters, there is not much to be said about this aspect of total cost—not because it represents a trivial investment, but because the differences in costs of installation among comparable models tend to be relatively small.

ANNUAL STORAGE COST (WITH ROOM TEMPERATURE AT 70°F)

Standard Gas Water Heater (40-Gallon)	Water Temperature	Heat Loss (Btu/day)	Dollars/Year*
	160°F	28,000	$74
	150°	25,000	66
	140°	22,000	59
	130°	19,000	51
	120°	16,000	42

Energy-Saver Gas Water Heater (40-Gallon)	Water Temperature	Heat Loss (Btu/day)	Dollars/Year*
	160°F	17,000	46
	150°	15,000	40
	140°	13,000	35
	130°	11,000	30
	120°	9,000	25

Standard Electric Water Heater (52-Gallon)	Water Temperature	Heat Loss (kwh/day)	Dollars/Year*
	160°F	5.5	159
	150°	4.9	142
	140°	4.3	126
	130°	3.7	107
	120°	3.1	90

Energy-Saver Electric Water Heater (52-Gallon)	Water Temperature	Heat Loss (kwh/day)	Dollars/Year*
	160°F	2.8	80
	150°	2.5	72
	140°	2.2	63
	130°	1.9	55
	120°	1.6	47

Gas at 56 cents per 100 cubic feet (and assuming 75 percent efficiency); electricity at 7.9 cents per kwh.

The differences become noteworthy, however, when you consider making a much more sizable investment—for a solar water-heating system, say—discussed in chapter 4 and later in this chapter. It's likely that with such equipment you would save more on operating costs than you would with a conventional system. But the initial investment should be viewed as any other in the energy arena: You should try to get the greatest return for your money. The worksheet at the end of this chapter, like the other worksheets in the book, can help you decide where you can spend your money most wisely.

CONVENTIONAL GAS AND ELECTRIC WATER HEATERS

The conventional gas or electric water heater basically is just an insulated tank with a heater—either a gas burner or electric heating elements.

Until 1980, the most commonly used insulation was fiberglass, but more and more new models use polyurethane foam.

Upright water heaters are, typically, cylinders 4 to 5 feet high and 1½ to 2 feet in diameter. Popular sizes are 40 gallons for gas models and 50 or 52 gallons for electric ones. (The latter need to store more water because they don't heat it as quickly.)

Water heaters fall into two broad categories—"standard" and "energy-saver." The major difference between the types is that an energy-saver is better insulated than a standard water heater.

Operating Cost

One way water heaters are characterized is by their rate of energy input—in Btu per hour or in watts. In theory, the higher the value, the more quickly the model can heat cold water. Gas models commonly are rated from about 30,000 to 50,000 Btu per hour. A typical electric model is rated at about 4,500 to 5,500 watts (the equivalent of about 15,400 Btu per hour).

Electric water heaters have higher total operating costs than gas ones. (See the comparison in the section entitled "Reviewing Hot-Water Economics.") But differences in operating costs among various electric models or among gas models tend to be slight. There are greater differences among units when it comes to purchase prices and lengths of warranties.

A standard water heater would cost about 18 percent more to run than an energy-saver model. The saving with energy-saver models would be almost entirely the result of reduced storage losses because of their improved insulation. There is little difference in the cost of heating between gas energy-savers and gas standard models: The same is true for electric models. You can't change the temperature of the cold water entering your house, but you should be aware that seasonal temperature differences will affect your operating cost. If you live in an area where winters are cold, water entering the

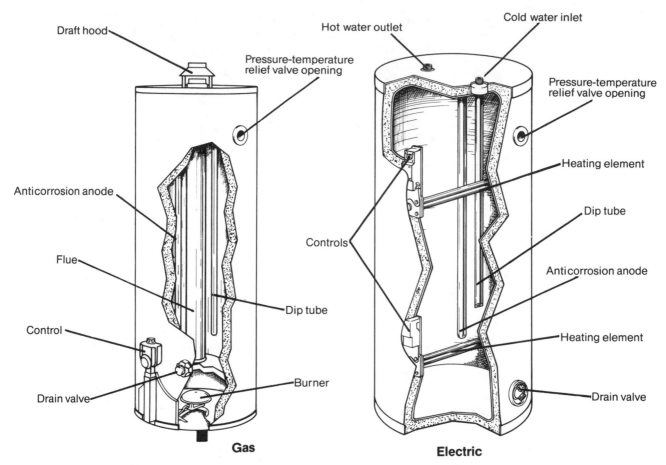

Figure 6.1 Construction features of typical water heaters

house will be at least several degrees lower during cold months and will require proportionately more heating (and thus more fuel) than when the weather is warm.

Hot-Water Delivery

Operating cost isn't the only important factor, of course. If you've ever soaped yourself in the shower and then suddenly found the water ice cold, you can appreciate the benefit of a water heater that can deliver plenty of hot water.

Manufacturers' claims for tank capacity may be optimistic. A "40-gallon" gas model may actually hold about 1 or 2 gallons fewer, and a "50-gallon" or "52-gallon" electric model might hold about 3 to 5 gallons fewer.

As water is drawn from the tank, there can be a fairly sharp stratification between the hot water still stored in the tank and the cold water entering the tank. (The cold water enters through a *dip tube,* which directs the water toward the bottom of the tank.) That layering means you can get most of the stored water out of the tank as hot rather than warm water. A gas model may deliver 29 to 33 gallons before the temperature at the hot-water outlet drops 20°F. Based on Consumers Union tests, an electric model with a larger tank delivers 35 to 42 gallons of hot water.

Because electric models tend to have a larger tank, they can hold more hot water than gas models. But when you finally run out of water, you have to wait much longer for an electric model to heat up a new tankful. Gas models produce hot water faster than electric ones. A slow gas-fired water heater could heat 37 gallons per hour, while the fastest electric may manage only 32 gallons per hour.

Features of Water Heaters

Most electric models have a heating element near the bottom of the tank, where the cold water enters, and a secondary element near the top, where the hot water is drawn off. After you use some hot water, the lower element comes on, but if you drain all the hot water from the tank, the lower element switches off and the top element comes on. That way, you don't have to wait for the entire tank to heat up before you have hot water.

Controls. Gas models have their thermostat and burner controls on the outside of the water heater, making adjustments convenient. Outside controls also make it easy to turn off the main gas burner and leave the pilot light on—something you might want to do before a vacation. Some models have a very low setting marked "vacation," but if the heater won't be exposed to freezing temperatures, you can save more by turning it off completely before leaving home for more than a day or two.

The controls on electric models tend to be more difficult to adjust. They may be hidden behind a panel that has to be unscrewed. For safety's sake, turn off the electricity before removing the panels, because the controls are connected to the electrical supply.

Thermostats on most electric models are marked in degrees Fahrenheit. But a few electrics and most gas heaters have thermostats marked only with words such as "hot," "normal," and "warm." Without degree markings, you need a thermometer to measure water temperature at the tap.

Heat traps. Normally, some hot water will circulate within the pipes, even if you're not using any water, because hot water normally tends to rise and cold water tends to fall. Such circulation turns the first few feet of water pipes into a radiator, giving off heat from the water. That's where a *heat trap* can be useful. A typical heat trap (see the accompanying illustration) is a loop of tubing mounted between the water heater and the water pipes to reduce the unwanted circulation.

Some heaters come with one or more heat traps. Others can accept heat traps but are not supplied with them.

Heat traps can contribute a worthwhile saving—about $9 per year for gas models and about $38 per year for electric models, based on 1990 average energy rates. If you buy a water heater not equipped with heat traps, you should add a pair of them when it is installed.

Drain valve. Drain off some hot water periodically to keep sediment from accumulating at the bottom of the tank; that will increase the life of the tank. In areas with hard water, it's best to do this every month. In areas with soft water, every three or four months is sufficient.

The best type of drain is the convenient brass drain valve, which looks like an outdoor faucet.

Some models have a drain valve with plastic threads for a hose connection. That type requires extra care when screwing on the hose to avoid cross-threading the fitting.

One type of undesirable drain valve is a threaded plastic pipe that passes through the center of a large knob that controls the drain valve. Turning the knob to open the valve makes the hose twist—and possibly kink. And if you turn the knob too far the wrong way when trying to close the valve, the valve stem may come out and spill the entire contents of the tank.

Corrosion protection. Water heaters typically have a "glass" lining in the tank to protect the metal from rust. As further protection, heaters also have at least one anode (a magnesium rod) suspended in the tank. Magnesium, a metal that corrodes readily, acts as a "sacrificial anode": Corrosion tends to attack the anode and spare the steel tank. A second anode probably would increase the life of the tank somewhat.

Warranty. Deluxe energy-saver models come with a 10-year warranty on the storage tank; others come with a five- or an eight-year warranty.

Does it make sense to pay more for a longer warranty? Water in some areas is unusually corrosive, and if you live in such an area, it might pay to purchase a water heater with a long warranty. However, warranties don't cover transportation or installation charges should the tank fail. And if you move from

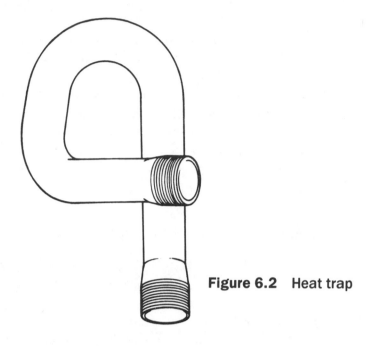

Figure 6.2 Heat trap

your home before the warranty expires, you would have paid the warranty premium on the tank for the next owner.

Safety

An electric heater is safe as long as it is wired properly. Gas models have a device that shuts off the fuel supply if the pilot flame goes out.

If the water temperature gets too high because of thermostat failure or other problems, both types of water heaters have an energy cutoff that shuts off the heat as well as a threaded hole for a pressure/temperature relief valve that can drain off some water.

Hydrogen, a highly flammable gas, is normally produced in water heaters—usually in small quantities—as a result of the reaction between the heater and an anticorrosion device. But hydrogen can build up when hot water isn't drawn for several weeks—while you're away on vacation, for instance. After a vacation, it's a good idea to run some hot water from a faucet before using a dishwasher or clothes washer; that allows the hydrogen to escape. Don't smoke or light a match near the faucet while it's venting.

The safety of a gas or electric water heater depends on its installation. Hire an experienced, licensed plumber or electrician to do the job.

Recommendations

If natural gas is piped into your home, you should buy a gas water heater, or a propane (LP) heater if there is no accessible natural gas supply. Unless you live in the shadow of a hydroelectric plant and enjoy an exceptionally low electricity rate, you're not likely to find a conventional electric water heater cheaper to run than a gas model. An energy-saver model appears to be worth its premium price over a standard model.

Cost aside, an electric model may present a drawback because it heats water more slowly once you use up all the heated water in the tank. That could be an important consideration if your family uses a lot of hot water.

INSULATION KITS FOR WATER HEATERS

If you can lower your home-heating bills by insulating the attic, it's only logical to assume that you could cut your water-heating bills by wrapping an extra layer of insulation around the tank.

A water heater can lose a considerable amount of heat while it's standing by. At 1990 average fuel rates, even a well-insulated energy-saver model storing 140°F water in a 70°F room over a year's time can accumulate about $62 worth of storage losses for an electric water heater or $32 for a gas one; older water heaters or standard models, which are less well insulated, could lose twice as much.

By adding insulation to the tank, you could reduce the amount of heat lost through the walls of the water heater, thus reducing the amount of energy required to keep the water in the tank hot.

A typical insulation kit consists of a fiberglass blanket, about 1½ inches thick, that wraps around the heater and is secured with tape. (The top of a gas-fired water heater should not be insulated; the insulation would interfere with the flue and create a safety hazard.)

In Consumers Union tests, a fiberglass kit reduced storage losses for a gas energy-saver water heater by about one-sixth, for a saving of up to about $7 per year at a natural gas rate of 56 cents per 100 cubic feet.

Any kit will produce a greater saving if installed on an older water heater that has relatively little insulation. The kit also will deliver a greater saving if your water heater is in a cold room or if its thermostat is set higher than 140°F.

It should be possible to install a kit in about an hour. The job is relatively easy, and the only tool needed may be a pair of scissors. When installing a kit, be sure to keep the insulation away from the heater's controls and wiring.

Fiberglass kits should carry a warning concerning the handling of fiberglass, as it is nasty stuff. When handling it, you should wear a dust mask, gloves, and a long-sleeved shirt. Later, wash your clothes separately so the fibers are not transferred to other clothing.

In choosing an insulation kit, first be sure it will fit your water heater.

The following two companies, among those whose products were tested for a Consumers Union report, were still in the business of manufacturing water-heater insulation kits as of late 1990:

- S & S Industries, Inc., P.O. Box 5538, Maryville, Tennessee 37802: Thermo Saver
- Thermwell Products Co., 150 East 7th Street, Paterson, New Jersey 07524: Frost King

TANKLESS ("INSTANTANEOUS") WATER HEATERS

Gas-fired, tankless water heaters are a familiar sight to those who have traveled abroad. They are about the size of a toilet tank and adorn the walls in many city apartments and country cottages. For years they have been used in Europe and Japan, but because the hot-water output of these foreign units is limited, they have never caught on in the United States.

Nevertheless, a few companies have been selling in the United States whole-house water-heating systems that work on the same principle. Tankless water heaters—also called *instantaneous heaters*—heat water as it's needed instead of letting gallons of heated water sit at the ready in a tank. The heart of the device is a network of copper tubing with metal fins like those on an automobile radiator, with a gas burner underneath. When you turn a hot-water tap to anything above a moderate flow, the burner automatically turns on, heating the water that passes through the tube. After a short delay—typically, 10 to 15 seconds—the water leaving the heater becomes hot. When the tap is turned off, the unit shuts down.

Gas-fired instantaneous heaters come in two versions. One runs on natural gas; the other uses propane (LP) gas. The two are virtually identical except for the sizes of the burner orifices, which are larger on the LP models to make up for propane's lower energy level. Electric models also are available, but none of those seen by Consumers Union can deliver water that's both hot enough and in sufficient volume to replace a conventional hot-water system.

Instantaneous gas water heaters come in various sizes. Certain small models, for example, are meant to heat water for only one sink tap, while the largest sizes are intended to replace a conventional water heater. These can heat water for a shower, a clothes washer, or a dishwasher, although they are not large enough to do more than one such task at a time. These whole-house models cost almost twice as much as a conventional water heater. The extra outlay theoretically is offset by lower running costs.

Reviewing Hot-Water Economics

An instantaneous water heater can cut energy costs primarily because, unlike a conventional water heater, it doesn't keep a tank of hot water standing all the time. The energy consumed in keeping that water hot is called a *storage loss.*

As we've noted, storage losses are not the huge waste some would have us believe. They generally account for about 25 to 30 percent of the cost of running a water heater, which typically is a couple of hundred dollars a year. The potential saving from eliminating storage loss, then, is likely to be about $100 per year for all but standard electric models, and it can be much less than that.

Instantaneous water heaters eliminate storage losses, but they don't eliminate all energy waste. They register a more-or-less-continuous loss from a burning pilot light, which, unlike the pilot on a conventional tank heater, provides no useful heating when the burner is off.

Any calculation of how much it costs to heat water depends on many variables—how much water is used, how much it's heated, the cost of the fuel, and so forth. Consumers Union engineers have formulated a model of use, based on a hypothetical family of four that lives where energy prices are the national average.

What follows later in this section is a comparison of the annual costs for heating 600 gallons of water per week. The hypothesis assumes that all family members bathe daily and wash their hands four times a day, and that a clothes washer is run twice a week. It is also assumed that the heater is raising 60°F water to 120°F.

The energy costs used are national averages, based on early 1990 rates: 56 cents per therm (about 100 cubic feet) of natural gas; 7.9 cents per kilowatt hour for electricity; 76 cents per gallon for LP gas; and 88 cents per gallon for oil. (The exact dollar figure given for each appliance would vary with energy costs, of course, but the heater's relative fuel efficiency would remain the same.)

As the figures show, our hypothetical family would save only a small amount annually by replacing a conventional gas water heater with an instantaneous model. Ultimately, a tankless unit might justify its higher price tag if it will replace a worn-out heater, but it's not a sensible replacement for a functioning one.

ESTIMATED ANNUAL COST FOR HEATING 600 GALLONS/WEEK

If natural gas Is available:	Cost per Year
Instantaneous water heater (gas version)	$131
"Standard" gas water heater	158
"High-efficiency" gas water heater	140

If natural gas is not available:	
Instantaneous water heater (LP version)	206
"High-efficiency" LP gas water heater	211
"Standard" electric water heater	448
"High-efficiency" electric water heater	409
Oil-fired water heater	147

However, as we've noted, if you have an electric water heater, switching to a gas model—any kind, whether natural gas or propane, instantaneous or conventional—offers dramatic savings. Any electric water heater—even a modern, high-efficiency one—would cost at least twice as much to run per year as a gas water heater. (The cost difference diminishes, of course, if your local gas prices are higher or your electricity rates are lower than those used in the Consumers Union calculations. And in some communities, electric utilities offer "off-peak" rates that may make the cost of heating water with electricity less punitive.)

While your saving would be greater with natural gas, even the more expensive propane has a vast advantage over electricity. Remember that it's the switch in *fuel* that makes the major difference—the instantaneous gas heater has only a minimal advantage over conventional gas-fired designs.

How Much Hot Water?

Manufacturers of instantaneous heaters claim that they provide endless hot water, and Consumers Union's tests showed that to be true. An instantaneous water heater won't run out of hot water the way a conventional water heater does. But that's a convenience feature, not a money-saver. If access to a bottomless well of hot water encourages you to use more of it, your water-heating bill will go up, not down.

On the other hand, an instantaneous heater might force you to use less hot water than you'd like—simply because it can't deliver as much hot water per minute as a conventional water heater. A regular water heater can deliver 30 to 40 gallons of, say, 120°F water as fast as your house's plumbing will handle it. Some instantaneous heaters can deliver at about the same maximum rate as a regular water heater—but the water won't be very warm. Others have internal restrictors that reduce the maximum flow to ensure delivery of reasonably hot water when the tap is opened fully.

Temperature rise at a given flow rate is the figure by which one instantaneous hot water heater can be measured against another. That's because the temperature of the water delivered at the outlet depends on the temperature of the water supplied at the inlet.

While a storage-type heater set for, say, 120°F will deliver 120° water whether the inlet water is 45° or 70°, an instantaneous heater with a 60° temperature rise will deliver water at 105° or 130°F under the same conditions.

With almost any instantaneous water heater, it may take longer to run a hot bath or fill a clothes washer. And a flow-reducing shower head is all but mandatory. Also, the instantaneous heater won't deliver hot water at a very low flow—an inconvenience when you want only some warm water to wash your hands.

In addition, you may not be able to use a dishwasher effectively with one of these tankless heaters. Many dishwashers (for example, older models that don't have a built-in booster heater) require a 140°F water supply to function effectively. Some instantaneous heaters may fall short of supplying water that hot.

Safety considerations preclude installing an instantaneous heater in a closed space, such as a closet, bedroom, or bathroom. Like a regular gas water heater, an instantaneous heater needs a gas hookup and venting to the outside.

Recommendations

A whole-house instantaneous water heater is a good alternative to a conventional gas-fired heater only for those with special needs: for people who don't have room for a regular water heater, for example, or for shower addicts who want an endless if somewhat low-volume flow of hot water.

As for durability, an instantaneous model should last at least as long as a conventional water heater—10 years or so. In areas where the water has a high mineral content, it might even outlast a conventional water heater, because instantaneous heaters are not as prone to mineral buildup as the tank types.

SOLAR WATER HEATERS

Solar energy reaches the earth at an average rate of some 100 to 300 Btu per square foot per hour, which provides plenty of direct sun power. Unfortunately, there's a big difference between having the energy there and actually being able to use it.

In the broadest sense, virtually all the energy we harness is solar. Windmills, hydroelectric generators, wood-burning stoves—all derive their energy content from the sun, as do fossil fuels such as gas, oil, and coal. Popularly, however, solar energy has come to mean a more immediate utilization of the energy in the sun's rays. Solar water heating seems to be the most practical of the direct uses of solar energy for consumer use at this time.

There are several reasons for this. First, converting solar radiation to heat (instead of to other forms of energy) is a simple, low-technology task—as can be appreciated by anyone who has ever opened the door of a car that has been parked in the sun for a while. (A water-heating system, of course, is more complex.) And because the need for hot water (unlike home heating) is largely independent of climate, good use can be made of sunlight even in the South, where concern about the cost of home heating tends to be low. This year-round demand for hot water means that it's possible to amortize on a year-round basis the investment in collectors, storage tanks, and auxiliary equipment. And the temperature range required for household hot water (100° to 140°F) is low enough to be attainable with relatively simple solar collectors, yet high enough to permit a reasonable quantity of energy to be stored in a not-too-bulky tank.

While the purchase and installation costs of a solar water-heating system are higher than for a conventional water heater, there is potential for recovering the investment through lower energy bills. Furthermore, a solar system represents a possible hedge against inflation. The prices of conventional fuels may rise, but solar energy is free.

Solar heaters have been around for a long time. At the turn of the century, solar-energy equipment very similar to the equipment being sold nowadays provided much of the hot water for homes in the South and the Southwest. In 1892, one company offered the Climax Solar-Water Heater ("sufficient for 3 to 8 baths") for $15. In Florida in the 1940s, solar water heaters outsold conventional gas models by two to one.

By the early 1950s, however, the relative cheapness of electricity and natural gas had eliminated the United States market for solar water heaters. But since the 1970s, with conventional forms of energy becoming more and more expensive and often scarce, solar energy has made a comeback.

Solar water-heating systems cost less than most other kinds of solar hardware, and they can be used throughout the year in nearly every region of the country.

How Solar Water Heaters Work

A solar water heater consists of *collectors, a storage tank,* and some means of transferring the energy from the collector to the water in the tank. A backup heater generally is provided to supply hot water when the sun's energy isn't adequate to meet the demand. And some sort of electronic *controller* prevents the hot water stored in the tank from being transferred back to a cold collector when the sun isn't shining.

Collectors. The collectors in a solar water heater are straightforward devices. In their simplest form, they consist of a set of water-filled tubes that are exposed to the sun. (A length of garden hose lying in the sunshine would be a simple version of a collector.) Dark colors absorb solar radiation well, so the surfaces of the tubes are often blackened. The tubes usually are fastened to a blackened metal plate that provides further absorptive area and conducts heat to the water in the tubes. Water circulated through the tubes becomes heated and fills the storage tank.

In a very warm climate, that's about all that's needed to operate a solar water heater. If a collector is to operate effectively in a cold climate, however, something more elaborate is required: A good deal of heat from the water would be lost to cold, outside air unless the collector's back and side surfaces were insulated. And heat would be lost from the front surface unless some sort of transparent glazing were used that would pass short-wavelength solar radiation while blocking longer-wavelength reradiation from the collector. If very high temperatures are required, double glazing may be used, but the extra cost of this approach seldom is justified for heating water.

Climates that expose the collectors to freezing temperatures require additional precautions and thus add to the costs. You can, of course, circulate antifreeze solutions through the tubing. But this approach requires a recirculating antifreeze loop (hence the name *closed-loop system*) to keep the antifreeze out of the house's hot-water supply. A heat exchanger is also needed with this system to transfer the heat from the antifreeze to the water that is to be heated. Alternatively, the piping to a collector can be designed so that the water in the tubes is drained out of the collector when it's not being heated (a *drain-down system*) to prevent freezing.

Both systems work; each has its own advantages and drawbacks. Closed-loop systems always recirculate the same antifreeze mixture, which eliminates the problem of mineral deposits in hard-water areas and allows corrosion inhibitors to be used. But closed-loop systems require an antifreeze compound that can remain stable for long periods in high temperatures. The antifreeze must also be low in toxicity, in case the heat exchanger leaks. Such antifreeze can be expensive. And the heat exchanger hinders somewhat the efficiency of a closed-loop system; a double-walled heat exchanger, required under some codes, hinders the efficiency even more.

Drain-down systems avoid these problems, but they are subject to scaling and clogging by water impurities. They also make somewhat heavier demands on the pump, because it's harder to pump water up to an empty collector in a high location than it is to circulate liquid in a filled loop.

Storage tank. Solar hot-water systems use a storage tank that is thermally insulated and lined with a corrosion-resistant material. Backup heating, usually in the form of electric heating elements, generally is provided in case the homeowner requires more hot water than the solar collectors can provide. (Electric backup heating is commonly used despite electricity's high cost per Btu—partially because gas- or oil-fired heaters have inherently higher storage losses than do electric heaters, which can be insulated better, and partially because it's simpler and easier to install electric heating.)

The size of the storage tank for a solar water heater is an important factor in the system's performance. If the tank is too small, it can't store enough water to take care of the household's needs. Once the water in the tank reaches its maximum temperature, no more water will be heated. A tank that's too large, however, might permit the collectors to collect energy as long as the sun is out, but the stored water might never become hot enough to be usable. It would be very expensive to use electric backup heaters to keep a large tank of water heated.

Controllers and other hardware. If hot water circulates through a collector when the sun isn't shining, the collector would act as a cooler. Because that's hardly what is wanted, a temperature-sensing controller is used; it switches on the circulating pump only when the water in the collector is hotter than the water in the tank. Some controllers are proportional—instead of cutting off flow completely when the temperature difference between the tank and the collector decreases to a preset value, they reduce the flow rate as the difference becomes less.

The *circulating pump* usually is a somewhat smaller version of the circulators used with home-heating boilers. The pump must simply circulate the water at some desired rate of flow, do it without using too much electricity,

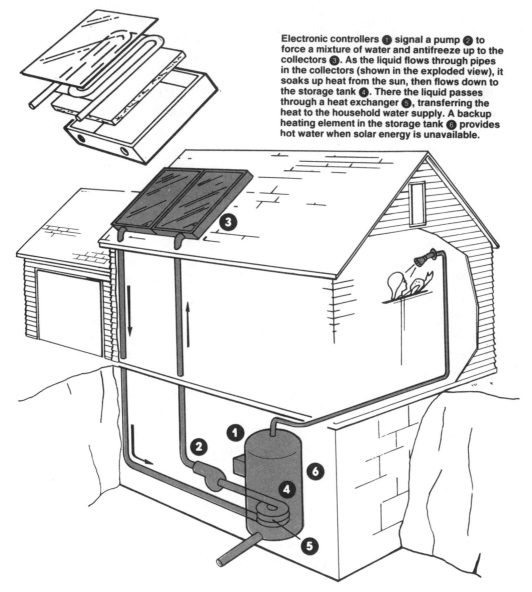

Electronic controllers ❶ signal a pump ❷ to force a mixture of water and antifreeze up to the collectors ❸. As the liquid flows through pipes in the collectors (shown in the exploded view), it soaks up heat from the sun, then flows down to the storage tank ❹. There the liquid passes through a heat exchanger ❺, transferring the heat to the household water supply. A backup heating element in the storage tank ❻ provides hot water when solar energy is unavailable.

Figure 6.3   A typical solar water heater

and require little or no maintenance. In a drain-down system, the pump must have sufficient reserve capacity to push the water up to the collectors each time the system goes on. Because untreated water can be quite corrosive, a pump for a drain-down system must have greater corrosion resistance than a pump in a closed-loop system, which handles antifreeze with added corrosion inhibitors.

Pipes connecting the various components of the water system are insulated to minimize heat loss. Tempering valves frequently are provided; they control thermostatically the outlet water temperature by mixing the hot water from the tank with some cold tap water. The collectors must be secured to whatever location is selected for them. Mounting hardware, while not strictly part of the water heater, is nevertheless important if the system is to operate for many years without trouble.

Performance

In the mid-1970s, sales of solar-energy hardware for water heating seemed to be at an all-time peak, although the technology was still evolving. Consumers Union tested five systems, each sold as a complete, brand-name package. The systems were ordered for use in the New York City area and specified for a family of three "using a lot of hot water." (That wasn't a precise specification, but it's one that a homeowner might use.)

What Consumers Union's engineers received were systems in which both the collectors and the storage tanks varied considerably in size. The smallest system consisted of a 65-gallon tank and 38 square feet of collectors. Two others each had about 50 square feet of collectors, but one of these came with an 80-gallon tank and the other with a 120-gallon tank. Yet another system had even more collector surface—64 square feet—combined with an 80-gallon tank. And the last system matched 69 square feet of collectors with a 120-gallon tank.

The tests provided insights into the performance of the equipment under worst-case conditions—short, sometimes cloudy winter days—as well as long, sunny summer days. The systems were run long enough to determine with some precision what their annual water-heating capabilities were.

The equipment was connected to a system of valves controlled by a timer, so that hot water could be drawn according to a schedule that approximated our hypothetical family's pattern of use. The electric backup heaters in the storage tanks were not used. Whenever water was drawn from the system, the amount and temperature of water used was recorded, and the amount of solar energy transferred to the water was calculated. The collectors themselves seldom heated the water to 140°F, except during the summer and early fall. Because 140° is the temperature specified for nearly all dishwashers, the amount of backup heating that would be required to achieve that temperature was computed. That, in turn, allowed a determination of the percentage of the family's hot-water needs supplied by solar.

As could be expected, the variations in collector size and efficiency, combined with the different tank capacities, resulted in considerable differences in performance. For instance, on two sunny December days, the best-performing system supplied about 55 percent of the hot water for the hypothetical family. The smallest model supplied 25 to 35 percent; another system, 31 percent.

Over a two-week midwinter period, which included both clear and cloudy days, the most efficient model provided 36 percent of the hypothetical family's hot water; the least efficient one contributed 16 percent. During summer days, any of the models supplied from 90 to 100 percent of the hot-water needs.

The specific numbers were not very important, but they indicated that a well-designed and well-constructed system, operating properly, could provide virtually all of a family's hot water during the summer and 50 to 75 percent of it annually. A backup system would be needed in the winter and on cloudy days.

Installation

Once a solar water heater was installed, it was hard to tell whether or not it was working: As long as the backup heating element was on, the system provided hot water. As a result, it wasn't possible to tell whether the solar collectors or the backup was doing the heating—at least not until the next utility bill came in the mail. This means that a faulty controller or a malfunctioning pump could put the solar collectors out of commission entirely, but the homeowner might not know it.

To clarify this, Consumers Union testers added a flow indicator containing a "pinwheel" that spun when water flowed through the system; if the pinwheel spun at night, or failed to spin on a sunny day, that was a clear sign that something was wrong with the equipment. To help spot operating problems, we recommend that manufacturers or installers include a sight glass or a visual flow indicator in every system so that homeowners will know when the equipment is working.

Many of the problems affecting solar water heaters have been the result of bungled installation. In one large study, for example, some valves were installed backward and some pipes were left uninsulated. In one case, a collector was mounted on the north side of a house—away from the sun. While few of the problems could be attributed to anything but a manufacturer's or installer's inexperience, they were serious nonetheless, and they led to a disappointing 21 percent saving of energy, on average, for the 100 systems in the study. Some of the systems saved nothing at all. However, once they were adjusted, repaired or replaced, the overall performance improved.

If you're in the market for solar, be sure to check an installer's qualifications thoroughly before you buy (see chapter 9). Because solar water heaters require periodic maintenance, make sure your contract spells out the cost and terms of servicing. Also, find out if the manufacturer supplies an owner's manual: If one is not available, you might want to consider another system.

Costs and Savings

The amount of heat a solar water-heating system supplies depends primarily on the square footage of the collectors exposed to the sun. If cost were no object, it would be a simple matter to rig one or more large storage tanks to a very large array of collectors and use the sun to heat all of a family's hot water. But designers and homeowners usually have to make trade-offs between cost and performance, settling for a system that can supply only part of the hot water needed each year.

From a technological standpoint, solar water heaters work. Properly built, installed, and maintained, they can provide hot water, in nearly every part of the country, at a significantly lower operating cost than other types of heaters. But when purchase and installation costs are considered, the picture may be less rosy.

Nevertheless, as the price of conventional energy increases, solar energy should become more attractive, although without some sort of subsidy, the initial purchase price is discouraging. And even if the cost is subsidized, a solar water heater still represents a sizable onetime investment.

Keep in mind that installing a solar water heater is not the only way to achieve a saving on heating water. The other approaches to conservation described in this chapter sometimes can yield as large a saving as a solar water heater can.

One important factor to consider is the way your household uses hot water. With a solar water heater, *when* you use hot water can be as important as *how much* you use. Let's say your family uses hot water mainly in the evening. Keep in mind that if you were to empty the tank of hot water, you couldn't get more until perhaps the middle of the following day, unless you used electric backup heaters. And if you did depend on them to heat more water, you would be heating that water with expensive electricity, not solar energy.

Similarly, if you insist on doing all your laundry every Monday, rain or shine, you might again find yourself heating much of the water with electric backup heaters. This is another example of how your saving with a solar heater would then be much less than you might otherwise expect. On the other hand, if you disconnected the backup heaters, you would reduce your energy bill for water heating to virtually zero. But you also could expect to run out of hot water much more often.

The cost of the equipment, of course, is a major factor when considering a solar water heater. Equally important, however, is the amount you now spend to operate a conventional water heater; the amount you could save by switching to solar will depend on this current figure. The Water-Heating Worksheet in this chapter can help you estimate your current costs and the potential yearly saving likely with alternative water-heating methods, including solar.

As a guide through the steps, we include sample calculations for our hypothetical family. We assume that the family now has a standard electric water heater and pays the 1990 national average electricity rate of 7.9 cents per kwh. We also assume that the family is considering a $3,000 solar water heater.

The example shows that the family now spends about $544 per year for hot water. The annual cost of installing and operating the solar water heater comes to $442 after taking into account the following: the annual equipment cost (the cost of the equipment divided by its expected life of 20 years), the amount of electricity needed for backup heat, and a rough estimate of annual maintenance costs.

So, for the hypothetical family, the estimated annual saving with solar would be roughly $102—not a very dramatic yield. Note, however, that our calculations do not take into account any increases in the cost of conventional energy. Obviously, a change in the price of conventional energy would make a substantial difference in the saving, as the following examples show.

Assume that the hypothetical family was paying 14 cents per kwh for electricity. At that rate, the family would pay $964 per year to operate its conventional water heater. The annual cost of operating a solar water heater would also increase—to about $522, because the backup heat would cost more. The annual saving, however, would jump from $102 to about $504, which would make a solar water heater much more attractive from a purely economic point of view.

Suppose, on the other hand, that the family had a standard gas-fired water heater. Because gas is a less expensive source of energy than electricity, the annual cost of hot water would be only $191—at a rate of 56 cents per 100 cubic feet. Switching from gas to solar would save the family only about $100 a year.

Recommendations

Homeowners in many parts of the country do not have a strong economic incentive to invest in solar heating because electricity, oil, and gas are still relatively inexpensive. If you are considering the change, however, remember that solar hot-water systems are feasible from a technological point of view, but the systems purchased must be dependable and durable, as well as properly installed and maintained.

Keep in mind, too, that solar isn't necessarily the most cost-effective way to reduce the price of heating water. Some of the alternative ways described in this chapter may be more attractive to you.

HOW TO CALCULATE YOUR WATER-HEATING COSTS

The worksheet that follows can help you calculate what you pay for hot water each year and can help guide you in deciding how best to reduce that cost. By following the steps explained below, you can determine the approximate annual net cost for a number of alternative methods for heating water, based on the amount you would have to spend each year for the estimated cost of operation and maintenance, over the expected life of the system.

If you already own a gas water heater, it probably will not pay you in the near future to convert to another type of water-heating system. And if you already have invested in a solar system or a heat pump, it likewise probably won't pay you to change your water heater now. That's why this worksheet addresses only the possible saving of changing over from a conventional electric water heater—standard or energy-saver. To guide you, we include an example showing the calculations for a hypothetical family of three weighing the pros and cons of replacing a standard electric water heater with a solar hot-water system.

The figures you derive won't be precise because the worksheet includes some estimates and assumptions necessary for simplicity's sake. Still, the results should be accurate enough to help you select the most effective way to cut energy costs for hot water.

The worksheet does not take into account increases in energy costs or the rate of inflation. As a result, the saving you calculate will be conservative. You can, however, work out costs and savings at different energy prices. In the worksheet, the saving for the hypothetical family is calculated at an electricity rate of 7.9 cents per kwh—the 1990 national average.

WATER-HEATING WORKSHEET

I. How Much Hot Water Do You Use?

Estimate your weekly consumption. Record the number of different uses of hot water per week, as shown on lines 1 through 7. Multiply each by the number of gallons shown. (The figures given represent typical quantities of hot water for each type of use.) Add the subtotals and enter the result on line 8. (The hypothetical family uses 504 gallons of hot water a week.)

II. How Much Does It Cost to Heat the Water?

Determine the annual cost of heating that water. Multiply the number on line 8 by 34.5. (The number 34.5 represents the "heat units" or thousands of Btu needed each year to heat 1 gallon of water per week from 60° to 140°F.) Enter the result on line 9.

Enter the fuel factor. This is a number that takes into account the different amounts of heat produced by different fuels. The factor for electricity is .29; for natural gas, .013. Enter the appropriate fuel factor on line 10. (The hypothetical family has a standard electric water heater, so the entry on line 10 is .29.)

Enter the cost of fuel. Check your monthly utility bill to determine the price you pay for a basic unit of fuel—a kilowatt-hour (kwh) of electricity or 100 cubic feet (about 1 therm) of natural gas. Enter your fuel costs, in cents, on line 11. (The hypothetical family pays 7.9 cents per kwh—entered as 7.9.)

Calculate your cost per heat unit (in cents). Multiply line 10 by line 11. Enter the result on line 12 and on line 15.

Calculate your annual cost of heating water. Multiply line 12 by line 9 and divide by 100. Enter the result, rounded to the nearest dollar, on line 13. (The family in the example spends $398 each year to heat hot water, exclusive of storage costs.)

III. How Much Does It Cost to Store the Hot Water?

Enter on line 14 the number of heat units per year it costs to store hot water. If you have a standard electric water heater, enter 5,400; if you have a standard gas water heater, enter 8,000. For energy-saver models, enter 2,700 (for electric) or 4,700 (for gas). (The hypothetical family has a standard electric water heater, so 5,400 is entered on line 14 in the example.)

Calculate your annual cost of storing hot water. Multiply line 14 by line 15 and divide by 100. Enter the result, rounded to the nearest dollar, on line 16. (The family's annual cost of storing hot water is $124.)

IV. What Is Your Total Hot-Water Cost per Year?

Add lines 13 and 16. Enter the result on line 17. That number is your estimated total cost of hot water for one year. (For the hypothetical family, that cost is $522.)

V. What Will New Energy-Saving Equipment Cost?

Begin with the cost of the equipment. Determine the total cost, including installation and any auxiliary equipment and building modifications required. Enter the total on line 18. (The example shows $3,000 for installation of a solar system.)

Estimate the useful life of the new equipment. Warranties could be a guide to estimated useful life of equipment. Enter the estimated life of equipment (in years) on line 19. (In the example, we assume a 20 year equipment life for a solar water heater.)

Determine the annualized net equipment cost. Divide line 18 by line 19. Enter the result on line 20.

Estimate the cost of providing hot water with a new system. If the hypothetical family switched from its standard electric water heater to a solar water heater, its estimated annual cost of heating and storing water would be about 40 percent of the cost of using its present water heater.

Accordingly, in the example, the entry on line 21 is $293—40 percent of line 17, or $218 plus an estimated $75 a year for maintenance of the solar water heater. If the family installed a heat-pump water heater on its existing water-heater tank, the annual cost of hot water would be about 50 percent of line 17. (With heat-pump water heaters, maintenance costs should be minimal.) Although replacing a standard electric water heater with a conventional energy-saver electric model would not affect appreciably the cost of *heating* water (line 13), it would substantially reduce the cost of *storing* hot water (line 16). Recalculating lines 14, 15, and 16 for an energy-saver electric water heater would give a cost of $62 a year for storing hot water. That would make the total annual cost of heating and storing water $482, which would then be entered on line 17.

What if the hypothetical family had started out with an energy-saver model of a conventional electric water heater (instead of a standard model)? Then, if the family switched to a solar water heater, the estimated cost would be about 45 percent of its present cost, plus an estimated $50 a year for maintenance. If the hypothetical family installed a heat-pump water heater, mounted on that well-insulated energy-saver tank, its annual cost of hot water would be reduced by about 50 percent.

Enter your result—the annual cost of providing hot water with the new system—on line 21.

Add lines 20 and 21. This number represents the total annualized cost of purchasing and operating your new equipment. Enter this number on line 22.

VI. What Can the New Equipment Save?

Subtract line 22 from line 17. This number represents the saving obtainable with the new equipment (the example is based on 1990 energy rates). Enter the number on line 23. (*Note:* If line 22 is *larger* than line 17, the resulting figure represents a net loss if you install new equipment.)

WATER-HEATING WORKSHEET

I. How Much Hot Water Do You Use?

	Example			Your House		
	No. of Uses	Gal. per Use		No. of Uses	Gal. per Use	
1. Bath/shower	14	× 15 =	210	___	× 15 =	___
2. Laundry (hot)	7	× 25 =	175	___	× 25 =	___
3. Laundry (warm)	0	× 15 =	0	___	× 15 =	___
4. Dishwasher	7	× 11 =	77	___	× 15 =	___
5. Washing dishes by hand	0	× 4 =	0	___	× 4 =	___
6. Hand and face washing	21	× 2 =	42	___	× 2 =	___
7. Food preparations using hot water	0	× 3 =	0	___	× 3 =	___
8. Total weekly use (Add lines 1 through 7)			504			___

II. How Much Does It Cost to Heat the Water?

9. Heat units per year to heat water (line 8 × 34.5) — _17,388_ ——————
10. Fuel factor — _.29_ ——————
11. Cost of fuel, in cents — _7.9_ ——————
12. Cost per heat unit in cents (line 10 × line 11)
    Also enter this number in line 15) — _2.29_ ——————
13. Annual cost of heating water (line 9 × line 12
    ÷ 100) — $ _398_ $——————

III. How Much Does It Cost to Store the Hot Water?

14. Heat units per year to store water — _5,400_ ——————
15. Cost per heat unit in cents (from line 12) — _2.29_ ——————
16. Annual cost of storing hot water (line 14 × line
    15 ÷ 100) — $ _124_ $——————

IV. What is Your Total Hot-Water Cost Per Year?

17. Cost per year (line 13 + line 16) — $ _522_ $——————

V. What Will New Energy-Saving Equipment Cost?

18. Initial cost of equipment, installed — $ _3,000_ $——————
19. Estimated life of equipment (in years) — _20_ ——————
20. Annualized equipment cost (line 18 ÷ line 19) — $ _150_ $——————
21. Cost of hot water with new equipment — $ _293_ $——————
22. Estimated annual cost (line 20 + line 21) — $ _442_ $——————

VI. What Can the New Equipment Save?

23. Estimated annual saving (line 17 − line 22) — $ _80_ $——————

Water-Heating Recommendations

The most effective steps you can take to reduce your water-heating costs
will depend to a large extent on your individual circumstances. In fact, you
might find that you could do better—in terms of the saving achieved for the
money spent —by concentrating on reducing energy costs elsewhere in the
home.

Use the worksheet. The calculations required, while simple, may be tedious, but we think the worksheet is worth the time it will take to help you sort out your options. Before investing in new equipment, you may wish to consider some relatively low- or no-cost measures. If, for instance, your storage cost (line 16) is high, you might give priority to installing an insulating kit, as described earlier, or to lowering your heater's thermostat setting and experimenting with lower hot-water temperatures. If your annual cost for heating water (line 13) is very large, you might try cutting back on your consumption of hot water.

You may well find that your needs still are best met by a conventional water heater—either a standard one (with added insulation) or an energy-saver model.

LOW-FLOW SHOWER HEADS

Showers account for more than one-fifth of all the water used indoors, and more hot water than any other fixture or appliance. Installing a low-flow shower head is a simple and inexpensive way to cut down on that water use. For about $10 per shower installation, a typical family can save 5,000 to 10,000 gallons of water every year—plus the energy used to heat that water, perhaps $20 to $50. The saving depends, among other things, on how long and how often each person showers, which kind of shower head you currently own, and how much you pay for heating water.

The shower heads manufactured before 1980 could deliver 5 to 8 gallons of water per minute. Since then, nationwide water shortages have prompted public officials, consumers, and manufacturers to reduce that figure.

During drought emergencies, homeowners were motivated to install flow restrictors, washerlike disks that fit into a shower head's inlet end. But a flow restrictor can diminish the quality of the shower along with the flow, and many consumers simply removed them.

Meanwhile, manufacturers have gradually scaled back the flow rates of new shower heads while trying to maintain the quality of the flow itself. Shower heads designed for low flow usually have narrower spray area so that less water misses the person under the shower; they may also entrain air within the water stream in order to create a turbulent flow.

Most of the new shower heads available are low-flow heads that deliver 3 gallons per minute or less. New legislation in various states and in Congress is aimed at requiring all new shower heads to use no more than 2.5 gallons per minute.

Two major types of low-flow shower heads are available. One is the fixed-position head, the familiar swiveling kind that screws directly onto the shower pipe. The other is the hand-held "microphone" type, familiar in parts of the world where water, and especially hot water, is in short supply. These latter versions connect to the shower head via a flexible hose, or to a pipe fixture on the bathtub water outlet. (A new bathtub inlet must be installed to accommodate such a device.) They usually come with a clip or fixture that can position them like a fixed shower.

Designs and styles vary widely, too. The simplest are fixed-position heads that deliver only one type of spray, such as a steady rain or a mist. Others have adjustment knobs or levers that give two or more spray patterns, sometimes including a pulse or "massage" spray.

In a panel test conducted by Consumers Union, opinions and comments were obtained from the panelists after their showers in an effort to determine exactly what they liked or didn't like about each head.

No general agreement emerged about what made a good shower. Opinion was clear about more specific matters, however. Panelists generally were unimpressed with the number of features some heads offer. Nearly all of the testers strongly disliked the center-adjustment knobs that require reaching into the stream to adjust the flow pattern. Panelists who used one type of flow for lathering and another for rinsing, say, markedly preferred the type that can be adjusted on the side or periphery of the head.

Installation

Installing a new shower head is a simple task. Follow the directions that come with the head, which usually instruct you to remove the old shower head with large pliers or an adjustable wrench. If the old head doesn't loosen easily, steady the pipe inlet with a pipe wrench. If you are concerned about the bathroom plumbing's finish, you may want to use a wrench that avoids biting into the soft pipe metal and marring it or its plated finish. If the connections aren't very tight, you might be able to get by with placing a rag between the jaws of the pliers to minimize damage to a fixture's finish. Once the old head is removed, simply screw the new head into place, using the appropriate tool.

Unless the instructions specify otherwise, it usually is a good idea to wind the pipe threads with a few turns of pipe-joint tape, available at hardware stores.

Safety

A few manufacturers of low-flow shower heads have cautioned that their products have the potential for scalding the user unless the plumbing is fitted with special antiscald devices. The theory is that if someone flushes a toilet connected to the same cold-water line as the shower, a low-flow shower head will exacerbate the effects of any resulting pressure drop. With a low-flow head in place, flushing a toilet can cause a significant pressure change. Hot water forces the shower's cold water back up into its supply pipe, leaving only hot water coming out of the shower.

The problem shouldn't occur in bathrooms that are properly designed. Good plumbing practice specifies that ¾-inch supply pipes should serve bathrooms, and these would tend to minimize the problem. Further, major national plumbing codes call for the installation of thermostatic mixing valves, pressure-balancing valves, or antiscald valves in showers—any one of which should prevent the problem.

Check the pressure in your own bathroom by running the shower on your hand and have someone flush the toilet. If there is a significant temperature rise in your shower with your present shower head, chances are you would risk a scald using a low-flow head. Have an antiscald valve installed (a job for a plumber) or set your water heater lower (120°F or thereabouts) (or make sure the door is locked when you shower).

For Ratings of low-flow shower heads, see Appendix B.

SEVEN

Keeping Cool

To keep cool during hot weather, you can close the doors and windows and rely on air conditioning—if you're willing to pay for an air conditioner and the cost of the electricity it demands. There are ways to reduce the cost of cooling a house, but to do so effectively takes some ingenuity.

To stay cool—and save some energy and money in the process—you need to work *with* the weather, not *against* it. In brief, working with the weather involves four strategies: First, minimize the amount of heat generated within the house. Second, shade the house to reduce the impact of the sun's rays. Third, when the outside temperature is high, close the house to slow down the flow of warm air into the house from outdoors. Fourth, when the outside temperature drops, open the house to allow the house to cool.

Working with the weather doesn't mean you have to abandon air conditioning. In many areas, in fact, air conditioning is a virtual necessity. You can reduce an air conditioner's operating costs, however, if you follow our suggestions about working with the weather. You will also help your pocketbook if you choose a "high-efficiency" model, which can wring extra cooling out of the electricity the machine uses. The section in this chapter on "Room Air Conditioners" also includes two worksheets: one to help you choose a size that's right for the area the appliance is to serve and a second to help you calculate the cost of operating a room air conditioner. (These worksheets are not to be used for evaluating central air conditioning.)

In areas where nights are cool and humidity is low, a fan may be sufficient to provide cooling—either a whole-house fan or one or more window-mounted fans. Fans, of course, cost far less to operate than air conditioners. To help you evaluate fans, there's a section on the subject, including information about whole-house fans, window and box fans (for smaller-scale cooling and as possible alternatives to an expensive attic-fan installation), as well as guidance for ceiling fans, oscillating fans, and high-velocity fans.

INSULATION AND AIR CONDITIONING

Insulation helps slow the flow of heat from the house and is highly effective in lowering winter fuel bills. But many homeowners believe that insulation can substantially reduce air-conditioning costs as well: After all, if insulation helps keep heat *in,* shouldn't it also help keep heat *out?* Unfortunately, it's not that simple.

Consumers Union engineers examined the effects of insulation on air-conditioning requirements, working out cooling needs for hypothetical houses and reviewing existing studies. The closer they looked, though, the more elusive any substantial saving became. Insulating the house may not cut air-conditioning costs. And even if it does, it may not be the most cost-effective way to do so.

Winter weather affects heating requirements quite differently from the way summer weather affects cooling requirements. It's relatively simple to estimate how much heat will be required to keep a house comfortable. For any given house, the major factor is the difference between indoor and outdoor temperatures. The R-value of insulation in the house, the type of heating system, the construction of the house, and a family's living habits also affect heating requirements.

The heating degree-day, as noted in chapter 1, is a useful index for measuring heating needs. Once you know the number of heating degree-days, you can calculate how much heat will be lost by conduction through the walls, floors, and ceilings of a house. By carrying the calculations further, it's possible to make a reasonably accurate estimate of the amount of heat that can be saved by adding insulation.

It's far more difficult to estimate how much energy it will take to air-condition a house for summer comfort than to estimate how much heat you'll need. In summer there is humidity to consider as well as temperature, and we have no adequate single index (such as the degree-day) that relates these two factors to energy use.

What Insulation Can Do

To understand what insulation can and cannot do to reduce air-conditioning costs, you need to understand four major factors that affect comfort and air-conditioning requirements.

1. Humidity. Insulation slows the flow of heat, and the rate of heat flow is determined only by temperature differences, not by humidity. If it's 85°F inside and 85° outside, there won't be any heat flow through the walls, regardless of the humidity. A room at 80°F with 90 percent relative humidity may feel as uncomfortable as a room that's 85° with only 50 percent relative humidity. In many parts of the country, people turn on their air conditioners to relieve discomfort caused by humidity.

2. *The average seasonal difference between indoor and outdoor temperatures.* Even in the warmest parts of the country, the average difference between indoor and outdoor temperatures over the course of a cooling season is not large—a fact that limits the effectiveness of insulation. While there may be quite a difference during the hottest part of a hot day, the difference is very small most of the time. The saving in cooling costs will depend on the seasonal average difference, not on occasional temperature extremes.

3. *Outdoor temperature shifts.* In many parts of the country, the temperature inside a house can often be higher than the outside temperature, especially where days are hot and nights are cool. Insulation will help keep heat out when it's hot outdoors (which helps); it also will help keep heat in when it's cooler outside (which hurts).

4. *Direct heat from the sun.* During the summer, the sun's radiant heat has an important effect on air-conditioning requirements. Solar radiation is one factor that can be controlled to some extent, as we shall explain, but insulation has little effect on direct radiation.

Estimating Energy Savings

It's possible to take data on outside temperature and other factors (such as solar radiation) and then calculate the amount of energy required to maintain a given indoor temperature. This approach often is used to determine how large a central air conditioner a house requires, and the calculations generally are made on a "worst-case" basis to ensure that the cooling equipment will be able to handle the hottest days. The calculations are cumbersome, though—and they will tell you how large the air conditioner ought to be, not how much energy the unit will consume annually.

Insulation most likely will have a significant impact on cooling costs only in the warmest regions of the country—parts of Alabama, Arizona, Arkansas, California, Florida, Georgia, Hawaii, Louisiana, Mississippi, Nevada, New Mexico, Oklahoma, South Carolina, and Texas. And even in these areas, there are ways to save on cooling costs that may be more effective than insulation.

Consider a hypothetical house in Houston, a city with about 3,000 cooling degree-days a year. Let's assume that the house has 666 square feet of floor area, 1,600 square feet of walls (exclusive of glass), and no insulation. Let's also assume the following: that the central air conditioner has an energy efficiency rating (EER) of 6 (for more about EER, see the discussion later in this chapter); that it would be set to keep the house at a constant 78°F; and that the windows would be kept closed at all times. If insulation with an R-value of 19 were installed in the attic, the house's cooling requirements would be lowered by about $70 per cooling season, according to Consumers Union's calculations.

In a second calculation, let's assume that the Houston house already has R-11 insulation in the attic. Under those conditions, an *additional* R-19 (for a total R-value of 30) would save only an additional $7 per season.

There is a way to achieve a much more impressive reduction in cooling cost without adding that insulation. Simply by raising the indoor temperature from 78° to 80°F, cooling requirements for the house in Houston would drop by one-fourth. Raising the thermostat setting reduces the air conditioner's running time. And, because of the relatively small average difference between indoor and outdoor temperatures, a change of a few degrees in the thermostat setting makes a large difference in cooling requirements. Lowering the house temperature from 78° to 75°F, however, would increase daily cooling requirements by more than one-third.

Farther north, in Washington, D.C. (a locality with only 1,000 cooling degree-days), the picture is drastically different. There, outdoor temperature shifts are greater than they are in Houston, and outdoor temperatures tend to drop well below indoor temperatures at night. As a result, an uninsulated house in the Washington area would heat rapidly during the day and cool rapidly at night. Insulating the house would limit the daytime heat gain, but it also would limit how much the house would cool down at night. Consequently, insulation probably wouldn't lower cooling bills; it probably would change only the hours when you used the air conditioner.

That's the theory, and it's been supported by a U.S. government study in which a rather large ranch house near Washington, D.C., was equipped with storm windows and insulation in the ceilings, walls, and floors. The house was unoccupied and the windows and doors were closed around the clock. Government engineers maintained a constant indoor temperature and monitored the house's energy requirements. The insulation served, in effect, as a two-way roadblock, slowing the flow of heat into and out of the house. The result: no energy saving during the cooling season.

Working with the Weather

Using nothing but central air conditioning is a simple—but expensive and wasteful—way to achieve comfort in the summer. There are reasonably simple, less expensive approaches you can take—especially if you are willing to work with the weather.

Controlling heat generated inside the house is a necessary first step. You should reduce the use of lights and appliances whenever possible and schedule dishwashing, clothes-drying, and other heat-producing chores for early in the morning or late at night, when it's cooler outside.

Next, find ways to reduce the effects of the sun. Sunlight strikes a house with greater intensity and for a longer period of time in the summer than in the winter, so heat builds up within the house far more than it would during the winter. The greatest heat gain occurs in the afternoon, when the sun shines directly into west-facing windows.

You can reduce considerably the need for air conditioning by blocking out as much sun as possible. Here are four ways to reduce the impact of the sun's direct heat through shading and ventilation:

1. Paved sidewalks and driveways reflect heat and glare into the house. They absorb heat during the day and radiate it at night. A green belt of lawns or shrubs planted next to the house will help reduce the effect of that heat on the house.

2. A dark-colored roof and walls can absorb twice as much heat as light-colored ones do. If your house needs a new roof or the siding painted, consider using light colors to improve the house's ability to reflect heat.

3. Unless there's some escape route, heat that builds up in the attic will find its way into the house. Be sure the attic is adequately ventilated—with vents in the gable ends, in the eaves, and/or along the ridge of the roof. Also open any attic windows.

4. Because of the angle of the sun in summer, the windows on the east and west sides of a house become particularly vulnerable to the sun's heat. You can reduce heat gain by installing awnings, trellises, or other shading devices to protect those windows. Awnings should have ventilation slits or an air space separating them from the house; otherwise, they will become heat traps. A barrier of deciduous trees (not evergreens, which block out the sun in the winter) can be helpful by shading the early morning and late afternoon sun. The trees also help block heat reflected by neighboring houses. Keeping interior shades or draperies drawn during the day also will help reduce the effects of solar radiation, but exterior shading is far more effective.

Recommendations

Blocking out the sun and controlling heat inside the house will help lower your cooling bills. But there are two additional steps you should take to achieve the greatest saving in an air-conditioned house.

First, raise the setting on your air conditioner's thermostat. If you increase the setting a few degrees, you will reduce the amount of time the air conditioner runs, thus lowering the cost of operation. Because of the relatively small average difference between indoor and outdoor temperatures, raising the thermostat setting can make a big difference.

Second, monitor the outdoor temperature. Whenever it drops below the indoor temperature, it will pay you to turn off the air conditioner and ventilate the house in other ways. Opening the windows will help the house cool; using a whole-house fan or a window or box fan (discussed later in this chapter) can increase the rate of cooling. This approach will work most effectively in areas where the differences between daytime and nighttime temperatures are the greatest.

Note that because insulation slows the flow of heat during the day, the demands on the air conditioner can be reduced. You'll see a substantial saving, however, only if you ventilate the house quickly on cool nights. In a study conducted several years ago in central New Jersey, residents who stopped running their air conditioners and opened the windows whenever the outside temperature dropped below 68°F saw a 15 percent saving in their electricity bills for those days. The saving you achieve will depend on several factors: the amount of insulation in the house, the temperature you maintain during the day, the amount of time you run the air conditioner, nighttime temperatures, and relative humidity. Of course, you can also reduce energy requirements in air conditioning by employing a room air conditioner to cool and dehumidify only the immediate area you are using.

To sum up: In most parts of the country, it's best to insulate for winter needs and consider any summertime saving to be a fringe benefit. If your house already has some insulation, adding more just for air-conditioning needs probably will yield only a small saving. In the warmest areas of the country, where houses may have no insulation, adding insulation can help reduce cooling costs. But instead of worrying about what amount of insulation to install, consider raising the air conditioner's thermostat setting and shading the house. You're likely to save more if you take these steps than if you add insulation.

If you do add insulation, add it in the hottest walls and ceilings, not under the floors. The earth beneath the house usually is far cooler than the air, so insulating the floor closes off a useful escape route for unwanted heat.

ROOM AIR CONDITIONERS

An air conditioner's energy efficiency rating is the ratio of the conditioner's cooling capacity in Btu per hour to the number of watts of electricity it uses.

All else being equal, the higher the EER, the lower the operating cost. Choosing an air conditioner with a high EER saves both energy and money. Many states, as well as the federal government, have mandated increasingly higher minimum EERs for room air conditioners, and for good reason. An 8,000-Btu unit with an EER of 9.5 will use about 5 percent less electricity than a comparable unit with an EER of 9.0. To narrow your selection to air conditioners with high EERs, check the energy-cost labels (see chapter 8) on the models displayed in stores. (You can also send $1 to the Association of Home Appliance Manufacturers, 20 North Wacker Drive, Chicago, Illinois 60606, and request their latest directory of certified room air conditioners.) By completing the Air-Conditioning Cost Calculator Worksheet later in this chapter, you can see what effects various EERs would have on your electric bill.

But greater efficiency can also mean a reduction in an air conditioner's ability to dehumidify the air. Manufacturers are able to improve their units' efficiency partly by designing the units' cooling coils to run "warmer." But that reduces the coils' ability to condense moisture from the air.

Here are some measures you can take to overcome this inherent problem with high-efficiency air conditioners:

■ Buy a unit that's properly sized in terms of its cooling capacity. A high-capacity air conditioner isn't necessarily better than a smaller one. An oversized unit will cycle on for only brief periods, so it won't sufficiently dehumidify the air. A unit that's too small, of course, won't cool well enough. The worksheet on page 157 can help you determine the size that's right. Consider only those machines whose rated Btu-per-hour output is within about 10 percent of your calculation.

■ Run the air conditioner at a low or medium fan speed. That will move the air across the cooling coils more slowly, enhancing dehumidification.

■ Don't set the controls to make the room cooler just to compensate for high humidity. Studies have shown that people are usually just as comfortable in a room at 77°F and 60 percent humidity as they are in a room at 75°F and 40 percent humidity.

Directing the Air

If your air conditioner must be installed in a corner window, it's important for it to have adjustable louvers that can be set to direct cool air out into the room. Adjustable louvers can also be important in any installation if you want to spot-cool an easy chair, for example, when you've just turned the machine on in a warm room.

Many air conditioners have a second set of louvers that control up-and-down air flow. These louvers are often fixed to blow upward at an angle of about 45 degrees. Some models allow an adjustment of up-and-down flow. If you want to mount an air conditioner high on the wall (as in some through-the-wall installations), be sure the louvers on the unit you select will let you direct cool air downward.

The Thermostat

Many air conditioners have a thermostat that lets you choose a temperature setting somewhere between "colder" and "warmer"; only a few units have a thermostat marked in degrees Fahrenheit. Still, the unit should maintain whatever setting you choose without allowing the room temperature to fluctuate noticeably.

An excellent thermostat should not allow a temperature swing of more than about 1.5° over or under the temperature at which you feel comfortable. In most cases, the best type of thermostat is electronic, and it is mounted in front of the unit's cooling coils, where it can monitor the temperatures of the room air entering the unit, before the air is cooled. A less accurate thermostat may be located behind the control panel, out of the main stream of incoming air, and therefore is less capable of sensing the room's temperature.

Features

Cooling settings. A "high cool," "low cool," or intermediate setting changes the fan speed, with little effect on cooling. Still, the control has its uses: A high fan speed can provide a little extra breeze; a low speed cuts running noise and reduces energy use slightly.

Fan-only setting. Such a setting lets you circulate the air without any cooling effect. When your room isn't warm enough to cool, but you still want some air movement, it stirs up the air.

Getting fresh air. Some air conditioners have "exhaust" or "ventilate" settings to remove room air or bring in outside air. These settings help to freshen room air, but they detract from maximum cooling, because the warm air they bring in imposes an extra load on the machine and increases the electric bill.

Audibility. Quiet operation indoors is a big plus. The degree of outdoor noise that the unit emits also can be important if it is to be installed over a patio or adjacent to a room that isn't air conditioned.

Filter. All air conditioners have a filter to keep the cooling coils from becoming clogged with dust and thus less effective. The best filter arrangement lets you slide the filter out without removing the front grille. Look for a filter mounted in a frame or grid; that kind is easy to clean.

Condenser. In time, the condenser fins facing outdoors will clog with dust and grime. (It happens faster in a city, where there's a lot of traffic.) The condenser is accessible only by sliding the air-conditioner chassis from its frame, or by removing the entire unit. Vacuum cleaning may be enough, but if the grime is gummy, it may be necessary to hire a special service (check the yellow pages) for steam-cleaning the condenser.

Timer. A few models come with a timer that can be set to turn on the unit shortly before you come home from work, for example, or to shut it off automatically in the middle of the night. You can also buy a separate, heavy-duty air-conditioner timer with an attachment cord and plug.

Power needs. Highly efficient air conditioners draw less power for the same cooling load than less-efficient units and usually can be run on any 15-amp household circuit. A less-efficient air conditioner might draw a heavier current and require an independent circuit because of the greater power demand. In some areas, the local electrical code may demand an independent circuit no matter what type of air conditioner is installed.

Be sure the circuit's wall outlet is of the three-pronged grounded type and the circuit is protected by a circuit breaker or time-delay fuse. Don't use a toaster, iron, or other high-wattage appliance on the same circuit while the air conditioner is running. If you must use an extension cord, use only the heavy-duty type sold specifically for air conditioners.

Installation

Even if you don't have your air conditioner installed professionally, you'll still want to get someone to help with the installation. A room air conditioner is heavy, weighing from 45 to well over 100 pounds, and it can be tricky to put into place. A unit with a slide-out chassis makes the job safer. First mount the relatively light outer cabinet in the window or wall, then slide in the heavy chassis. Most units that don't have a slide-out chassis have brackets and leveling screws to hold them in place. The levelers let you tilt the air conditioner slightly toward the outside to keep condensation from dripping on the floor inside.

An air conditioner dehumidifies as it cools, so it collects a considerable amount of water in the process. To minimize dripping from the outside of the air-conditioner cabinet, there usually is a shroud around the outside fan called a *slinger ring*. It hurls the water into the air conditioner's hot condenser coils, where most of it evaporates and is carried to the outdoors in the form of water vapor. The slinger ring has a second, beneficial effect: It helps cool the coils, thereby raising the unit's overall efficiency. Even with a slinger ring, though, some air conditioners are prone to drip on extremely humid days. Therefore, it's a good idea to mount the air conditioner where it won't drip onto an occupied space, such as an entrance or a patio.

If your winters are cold, an air conditioner sitting in the window can be a significant pathway for heat loss. The mounting panels at the machine's sides do not make an airtight seal and are poor insulators in any event. Furthermore, cold air may find its way right through the machine. One alternative is to remove the unit each autumn and install a storm window to conserve valuable heat. Another, easier job is to cover the inside of the air conditioner with a transparent, tightly-fitting plastic cover (available at hardware stores and home-supply centers) that seals the cover right up to the window frame around the air conditioner (either inside or outside).

Recommendations

Unless you're convinced that you need an air conditioner, you may want to consider some alternatives that may save money in the long run; see the sections later in this chapter on whole-house fans, window and box fans, smaller-size portable fans, and ceiling fans. If you decide to buy an air conditioner, allow yourself plenty of time to shop before the first hot day arrives. That way, you can check for the best price on a model with a high energy efficiency rating: The higher the EER, the more economically a unit can be operated.

Here's an adapted summary of the U.S. Department of Energy's recommendations for saving energy with air conditioning:

■ Set the air conditioner's thermostat as high as possible. A level of 78°F often is recommended as an indoor temperature that's both reasonably comfortable and energy-efficient. The higher the setting and the less difference between indoor and outdoor temperatures, the less outdoor hot air will flow into the building. If the 78°F setting raises your home temperature 6° (from 72° to 78°, for example), you should save between 12 and 47 percent in cooling costs, depending on where you live.

■ Don't set the air-conditioning thermostat at a colder setting than normal when you turn on an air conditioner. It will not cool faster, and it will cool to a lower temperature than you need and also use more energy.

■ Turn off a room air conditioner when you leave a room for several hours.

You'll use less energy cooling the room later than if you had left the unit running.

■ Consider using a separate fan with a room air conditioner to spread the cooled air farther without greatly increasing power use. But be sure the air conditioner has enough cooling ability to handle the extended space you want it to affect. Otherwise, using a fan will reduce the unit's ability to cool any area adequately.

■ Don't place a high-wattage lamp or a television set near an air-conditioning thermostat. The thermostat senses the heat from these appliances, and this could cause an air conditioner to run longer than necessary.

■ Keep out daytime sun with vertical louvers or awnings on the outsides of your windows, or draw draperies, blinds, and shades indoors.

■ Keep lights low or off. Electric lights generate heat and add to the load on an air conditioner.

■ Do your baking and use other heat-generating appliances as early as possible in the day and during late evening hours.

■ On cooler days and during cooler hours, open windows instead of using an air conditioner.

■ Consider turning off a gas furnace's pilot light in summer, but be sure to remember to light it (or have it lighted by a service technician) before the heating season begins.

For Ratings of room air conditioners, see Appendix B.

CALCULATING AN AIR CONDITIONER'S OPERATING COST

The first step in achieving air-conditioned comfort without spending more than necessary is to match the air conditioner's capacity to the room it will serve. The Cooling Load Estimate Form that appears later in this chapter will help you determine how large an air conditioner you need. Second, shop for a high-efficiency model. Look for the energy efficiency rating on the appliance label; as we've noted, the EER is the number of Btu the air conditioner delivers per watt of electricity. The higher the EER, the more efficient the air conditioner. Finally, use the Air-Conditioning Cost Calculator Worksheet to determine how much it will cost to run an air conditioner in your part of the country. By comparing operating costs for different brands, you'll be able to determine which ones will cost less in the long run even if they carry what may seem to be a high selling price.

Air-Conditioning Cost Calculator

1. Determine your cooling-season electricity rate. Enter your rate in column 1. In the example below, we've used the rate of 7.9 cents per kwh (entered as $0.079, not 7.9).

2. Contact your local electric utility to find out the average number of hours of cooling used in your area. Enter cooling hours in column 2. (Our example shows 700 hours.)

3. Multiply column 1 by column 2. Enter the result in column 3.

4. Refer to item 11 on the Cooling Load Estimate Form to find the total cooling load for the room you want to air-condition. Enter that number in column 4. (Our example shows 6,000 Btu.)

5. Multiply column 3 by column 4. Drop the last three digits from the result. Enter that number in column 5.

6. Determine the energy efficiency rating (EER) of the brand and model you are considering. That number may be on the appliance label; if it's not, you can calculate the EER by dividing the unit's wattage into its Btu output—Btu/watt = EER. Enter the EER in column 6. (As an example, we use an EER of 9.1.)

7. Divide column 5 by column 6. Enter the result in column 7. The result is the estimated cost per 700-hour cooling season, in dollars, to run the air conditioner.

COOLING WITH FANS

Air conditioning works very well, but air conditioners are expensive, energy-hungry appliances. Fans, particularly portable fans, not only are much cheaper to buy than air conditioners, but they also use far less electricity and therefore are cheaper to run.

Fans don't cool the air; they cool people by accelerating the evaporation of perspiration from the skin. In areas where the humidity isn't uncomfortably high and nights are cool, fans of one kind or another can help keep you quite comfortable. The trick is to choose the right kind of fan and to use it in the most effective way.

Whole-House Fans

What once were called attic fans now are referred to as whole-house fans. The term *whole-house* has come into use to distinguish these large fans from the small units—more properly called attic fans—that are intended for ventilating only an attic.

Whole-house fans, with blades typically 24 to 42 inches in diameter, can be mounted in a variety of ways (see Figure 7.1). They are not as inexpensive

AIR-CONDITIONING COST CALCULATOR

1 Local Electric Rate	2 Average Hours of Cooling	3 Seasonal Electric Cost	4 Cooling Load	5 Cost Factor (Drop Last 3 Digits)	6 Energy Efficiency Rating (EER)	7 Operating Cost for Cooling Season
$0.*079* ×	*700* =	*55.3* ×	*6,000* =	*331,800* ÷	*9.1* =	$*36.46*
$0.___ ×	___ =	___ ×	___ =	___ ÷	___ =	$___
$0.___ ×	___ =	___ ×	___ =	___ ÷	___ =	$___

WORKSHEET: COOLING LOAD ESTIMATE FORM: CALCULATING THE SIZE THAT'S RIGHT

This worksheet, adapted from one published by the Association of Home Appliance Manufacturers, can help you estimate how much cooling capacity you need.

Preliminaries

1. Measure the length of each wall in the room.
2. Determine the area (length × width, in feet) of the floor and the ceiling.
3. Measure the area (width × height, in feet) of each window.
4. Measure the width of all permanently open doors. Rooms connected by a door or archway more than 5 feet wide should be considered one area. Take measurements for both rooms.

Calculations

Multiply the appropriate measurement by the factor given. Use factors in parentheses if the air conditioner will be used only at night.

1. _____ × 300 (200) = ..._____ 1
   width of permanently open doors, ft.

2. _____ × 14 = ..._____ 2
   area of all windows, sq. ft. (multiply by 7, not 14, for double glass or block)

3. Use only the line that's appropriate for your house.
   Uninsulated ceiling, no space above: _____ × 19 (5) = ..._____ 3
   ceiling area, sq. ft.

   Uninsulated ceiling, attic above: _____ × 12 (7) = ..._____ 3
   ceiling area, sq. ft.

Insulated ceiling, no space above: _____ $\times 8$ (3) = ... _____ 3
ceiling area, sq. ft.

Insulated attic above: _____ $\times 5$ (4) = ... _____ 3
ceiling area, sq. ft.

Occupied space above: _____ $\times 3$ (3) = ... _____ 3
ceiling area, sq. ft.

4. Enter length of all walls, in feet, as directed, and multiply by appropriate factor. Consider walls shaded by adjacent buildings as facing north.

5. If floor is on ground or over basement, omit this step and go to step 6.

_____ $\times 3$ = ... _____ 5
floor area, sq. ft.

6. If air conditioner will be used only at night or if all windows in the room face north, omit this step and go to step 7. Otherwise, enter the area for each window on the appropriate line. Do the multiplication. Multiply factor by 0.5 for any window with glass block; by 0.8 for double glass or storm window. Enter only largest number.

7. **Subtotal.** Add lines 1 through 6. Enter sum here: .... _____ 7

8. Climate correction

_____ $\times$ _____ = ... _____ 8
figure from line 7   factor from map

9. _____ $\times 600$ = ... _____ 9
people in room (min. 2)

10. _____ $\times 3$ = ... _____ 10
wattage of all lights and appliances in room (not including air conditioner)

11. **Total cooling load.** Add lines 8 to 10. Enter sum:
This number tells you how many Btu of heat build up in the room each hour. The air conditioner's cooling capacity (Btu/hour) should nearly match the heat buildup you calculated. A difference of about 5 percent between the number you calculate and the air conditioner's capacity should not be significant.

Wall Facing	Uninsulated Frame or Masonry up to 8 in. Thick		Insulated Frame or Masonry over 8 in. Thick	
Outside, north _____	×30 (30)	or	×20 (20) = ... _____ 4	
Outside, other _____	×60 (30)	or	×30 (20) = ... _____ 4	
Inside _____	×30 (30)	or	×20 (20) = ... _____ 4	

Window Facing	Window Area		No Shades		Inside Shades	Outside Awnings
Northeast	_____	×	60	or ×	25	or × 20 = _____
East	_____	×	80	or ×	40	or × 25 = _____
Southeast	_____	×	75	or ×	30	or × 20 = _____
South	_____	×	75	or ×	35	or × 20 = _____
Southwest	_____	×	110	or ×	45	or × 30 = _____
West	_____	×	150	or ×	65	or × 45 = _____
Northwest	_____	×	120	or ×	50	or × 35 = _____

Largest number _____ 6

as their $200-to-$400 price tags might suggest, because installation and such accessories as louvers, shutters, and remote switches can easily double their purchase price. On the other hand, a big fan working under the right conditions can cool and ventilate an entire house for about the same energy cost as running an air conditioner in one room.

If you merely open your windows in the evening, your house won't cool off very quickly. If the outdoor temperature drops from, say, 85°F to 75° in two hours, it will take about four hours for the house to cool down that much. A whole-house fan, however, replaces indoor air every minute or two, and it would allow the house to cool off in little more than two hours. What's more, you might even feel cooler before a thermometer confirmed that the temperature actually had dropped. Some people feel cooler when air moves past them as slowly as 100 feet per minute. A whole-house fan can produce that mild breeze.

If you keep a house fan running at night and turn it off in the morning, the house will retain some coolness during the early part of the following day, provided you close windows and draw shades or drapes. If you also have air conditioning, using this technique can reduce the air-conditioning load and help reduce cooling costs. But don't run a whole-house fan and an air conditioner simultaneously; you'll only push expensive cool air outdoors.

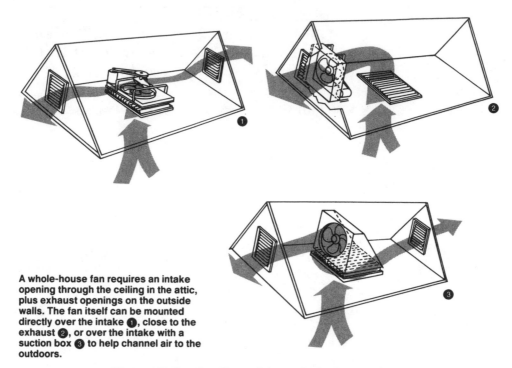

A whole-house fan requires an intake opening through the ceiling in the attic, plus exhaust openings on the outside walls. The fan itself can be mounted directly over the intake ❶, close to the exhaust ❷, or over the intake with a suction box ❸ to help channel air to the outdoors.

Figure 7.1   Cooling with a whole-house fan

Determining Correct Fan Size for Your House

The first step toward finding out what size whole-house fan to buy is to determine what air-moving capacity you need. Multiply length by width by height of each room, hallway, and stairwell to be ventilated. Don't include closets, pantries, storage rooms, or the attic. The total is the number of cubic feet of space you need to ventilate.

Use that total house volume to determine the air-moving capacity (expressed in cubic feet per minute, or cfm) that the fan should have. If summers in your area generally are hot, you should buy a fan that can change your house air completely every minute. A house with 6,000 cubic feet of living space, then, calls for a 6,000-cfm fan. Where summer temperatures are not extreme, a fan with a capacity of half the house volume should be sufficient.

Free air delivery is a term manufacturers use to rate their fans' air-moving ability when nothing restricts the airflow. That figure isn't very useful because a house's size, layout, and furnishings, as well as the fan's location and associated shutters, can all impede the air. Look for another, more useful rating that describes the fan's air delivery when working against a standard resistance (usually stated as 0.1 inch of water). If that rating isn't given, take 80 to 85 percent of the free-delivery rating as a reasonable approximation of the fan's real-life performance.

If the fan you decide on has only one or two speeds, then the house volume and fan capacity at maximum speed should match fairly closely. A variable-speed model can be a bit oversize, since you can run it at any speed you wish.

Installation

You'll need at least two openings for a whole-house fan: one from the living space to the attic and another from the attic to the outdoors. First see if you may be able to use existing openings: A door to the attic could serve as the air inlet if the fan is to be mounted in a gable end or through the roof, but you'd have to open the door whenever you used the fan. Attic windows could also double as exhaust vents, if you opened them every time you used the fan.

It's far more convenient to use fan shutters: these are hinged, horizontal slats that separate indoors from outdoors when the fan is off. They keep cold air out during the winter and they deter animals. When you turn on the fan, the shutters open. The simplest and cheapest fan shutters are designated as "automatic." They open solely by the fan's suction or air blast. But automatic shutters tend to reduce airflow more than shutters that open and close mechanically or electrically; the air reduction can be as great as 50 percent when the fan is turning at low speed. For some whole-house fan installations, manually operated or motorized shutters might be necessary, especially in an opening that isn't face-to-face with the fan. Such an opening may not have enough airflow "push" to let automatic shutters function properly.

It's best first to cut and frame the needed openings, and then fit them with shutters. The shutter opening nearest the fan usually should match the fan in size—a 3-by-3-foot opening for a 36-inch fan, for example. The other opening should be a bit larger. To figure its size, divide the fan's capacity, in cfm, by 750. (Air passing through these openings should not go faster than 750 feet per minute.) Figure an 8-square-foot opening for a 6,000-cfm fan, for example. That's the minimum unobstructed area that your fan would need. If the opening is to be screened (a good idea where insects are a problem), double that size.

Most whole-house fans draw only moderate power (350 to 700 watts), so they can be wired into an existing circuit. A few may impose a momentary start-up load of about 20 to 30 amps, but that surge probably wouldn't trip a circuit breaker unless the circuit already was heavily loaded. Fused circuits, however, should be protected by a slow-blow fuse. To help ensure electrical safety, you'll also need to ground the fan and any motorized shutters.

Safety

In case of fire, a fan can literally fan the flames. As protection, some manufacturers offer a fusible safety switch, either standard or as an option. In a fire, a fusible link melts, cutting off power to the fan and closing motorized or manually operated shutters. When a fan is installed, you may need to relocate your smoke detector. It should be out of the direct path of fan-blown air, but away from dead spots, too.

Shield the fan blades if people or animals can get near them. You can buy a frame-mounted blade guard or build your own. Shielding should not be impenetrable, however. Once a season, check the fan's drive belt and, if necessary, adjust its tension. Check whether the shutters are moving freely, and oil the fan and motor bearings, if the manufacturer's instructions advise that kind of maintenance.

Be sure the fan can't start up while you are servicing it. A fan that's not running is not necessarily a fan that is turned off; its overload protector or a thermostat may have merely stopped the motor temporarily. A manual on/off switch installed at the fan is your best safeguard; otherwise, shut off the power at the circuit breaker or fuse panel.

One more caution: Before turning on the fan, be sure that at least a couple of windows to the outside are open. Otherwise, the fan will create a partial vacuum that might extinguish pilot lights, cause a gas-fired water heater to malfunction, or even pull chimney soot into the house.

Recommendations

The expense and nuisance of having a contractor install a whole-house fan may seem forbidding. If so, there is an alternative: Buy several box or window fans, and use them instead. You may lose something in convenience, but you will save on carpentry. The following sections should help you decide whether you might be happier with such smaller fans than with a whole-house model.

Window Fans and Box Fans

A window fan is a scaled-down—and much cheaper—version of a whole-house fan. Window fans come in a variety of sizes, but 20-inch fans are particularly common. A typical 20-inch fan blowing out a window can move about 2000 cfm of air. That's enough to change indoor air once a minute in a two-room apartment. Window fans are designed specifically for mounting in a window; they don't come in a cabinet that would permit them to stand on a windowsill or on the floor. Box fans, on the other hand, are more versatile and portable; they can be built into a surrounding sheet-metal cabinet and can be placed anywhere you wish. Set in or near a window, they can ventilate. Set elsewhere in a room, they can be used for circulation.

Installation

To put a window fan in a double-hung window, just expand the fan's side panels and attach them to the frame with screws. Casement and sliding windows probably will need a special mount.

Some box fans offer optional window mounts—usually a pair of side panels held in place by a frame. The mounting frame generally has some sort of lip on which the fan hangs. A box fan used on a windowsill without side panels to seal the gaps along the sides is likely to lose about 10 percent in ventilating performance, when set to blow in.

If you'd rather not bother with side panels, you can put a box fan on a table 18 to 24 inches away from an open window. Mounted that way, it probably will exhaust air about as well as it would if mounted, well sealed, on a windowsill.

Other Features

Loudness. Window and box fans can be quite noisy at the highest speed setting. If noise is a potential problem (as in a bedroom), choose a multispeed model that runs quietly at its lowest speed.

Thermostat. A built-in thermostat saves you the trouble of shutting off the fan if the room becomes chilly. To set the thermostat, let the fan run continuously at its highest speed until the room cools down to the temperature you like. Then turn down the control slowly until the fan just goes off. The fan should then start up and shut down automatically in step with temperature fluctuations.

Fans equipped with a thermostat often have arbitrary numbers on the control to help you identify appropriate settings. Whenever you set the thermostat at night, remember to turn the fan off in the morning. Otherwise, the fan will restart when the temperature rises, drawing in hot daytime air if it's window-mounted. Some fans have an indicator light, a handy feature. It will remind you when the thermostat has turned the fan off.

Controls. The best location for controls on a box fan is the top, where they are easy to reach.

Ease of cleaning. Depending on the degree of air pollution where you live, you may want to clean the fan blade and other surfaces at least once a year.

Getting the grille off generally is a matter of removing a few screws, although you may come across a model with a pry-off grille.

Safety

It's always a good idea to disconnect a fan before you clean it, particularly one with a thermostat, which could restart the fan automatically while you're working on it. Fans with metal blades need a delicate touch when cleaning to avoid bending the blades and thus creating vibration when the fan is turned on.

Don't buy a fan unless it has a built-in circuit breaker to interrupt power if the motor overheats or draws excessive current.

Don't run a fan in a window while it's raining. If the fan should get wet, unplug it and let the fan dry thoroughly before using it.

Some window fans are exposed on their outdoor side; their blade guard covers only the indoor side. In a box fan, the blade is totally enclosed. Metal grillwork generally is quite protective. But be aware that some plastic guards may be flexible enough to spread and admit a curious child's fingers.

Recommendations

A window-mounted box or window fan can provide quite a bit of cooling, but the larger the space you're cooling, the longer it will take.

If you try to cool more than one room at a time (see accompanying illustration), note that the room in which the fan is exhausting will be the last to cool. So for quick, short-term cooling in a small area such as a single room, you might want to set a fan to pull in outside air, rather than exhaust. Such intake ventilation, however, isn't as efficient as the blowing-out kind.

In trying for longer-term cooling of more than one room, open at least one window in each room you want cooled. Try to open windows that provide the largest possible air sweep between the open windows and the one in which the fan is operating. But note that windows close to the fan should not be opened as wide as those farther away. Close all other windows, as well as all doors to rooms that don't need cooling. And be sure the fan is set to exhaust inside air.

Other things being equal, a box fan probably is a wiser choice than a window fan. Fitted with side panels and set in a window, the box unit can be just as unobtrusive and just as effective as a window-only model. And the box fan can be lifted easily from its window to serve as a circulator.

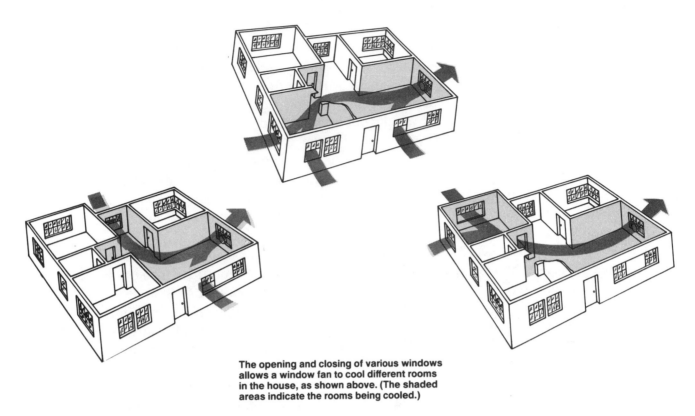

The opening and closing of various windows allows a window fan to cool different rooms in the house, as shown above. (The shaded areas indicate the rooms being cooled.)

Figure 7.2 Cooling with a window fan

Oscillating Fans and High-Velocity Fans

These fans can be used on a table or on the floor. They can be set to oscillate or to operate in a stationary position. On some models, the arc of oscillation may be adjustable. Certain oscillating fans have a safety feature in the form of a clutch mechanism that stops the oscillation if the fan strikes something in its movement from side to side. A fan without this feature can knock itself off its perch when it hits an unyielding obstacle. Any oscillating fan should have an adjustment so it can be tilted up or down. It also may have a provision for wall mounting.

A fan with a high-velocity circulator is designed to move air at higher speeds than possible with an oscillating fan. The extra air blast is obtained in part through the use of a shroud or cowling designed to channel the air into a relatively narrow, swiftly moving column. A high-velocity fan can be quite effective at stirring up stagnant room air.

Ceiling Fans

A ceiling fan is an effective way to stir up a breeze so that a room feels cooler. Such a fan's large blades allow it to move lots of air while running slowly and quietly. In addition, unlike a high-speed portable fan, a properly installed ceiling fan is physically out of reach, and thus safer if children are around. Be aware, however, that a ceiling fan isn't a substitute for an air conditioner, a whole-house fan, or the portable fans described here. A ceiling fan can't bring cool outside air into a room; it can only circulate the air that already is there. And you can't move a ceiling fan from room to room. Nor can you expect a fan—whether placed on the floor, in a window, or on the ceiling—to dehumidify; that's a job for an air conditioner.

Ceiling fans can be quite expensive, costing up to $250 and more. The less-expensive models generally have a plain painted motor housing, which might be appropriate for a kitchen or a room with contemporary or casual decor. The costlier models have a brass finish or baroque trim, both of which sometimes are more elaborate, or simply more noticeable, than you might want. You may prefer a ceiling fan designed for office or industrial use because of its relatively simple appearance.

The size of a ceiling fan is defined by the diameter, or sweep, of its blades. A typical diameter is 52 inches, but there are also 48-inch, and 36-to-38-inch models, suitable for a small room.

Manufacturers frequently describe their fans' performance according to the total amount of air moved in a given time, usually in terms of cubic feet per minute (cfm). Claims generally run from about 3,500 cfm for small models up to about 8,500 cfm for large ones. The most versatile fans have a continuously variable speed control, one that lets you run the fan at 50 revolutions per minute (or fewer) for a very gentle breeze, with the fan producing practically no noise at all, up to the fan's maximum speed. Next best is a three-speed control.

Ceiling fans often are promoted as winter energy-savers, moving warm air from ceiling level down to living level. But unless the fan is in a room with a high or vaulted ceiling or in a room heated by a wood stove, a fan's breeze is likely to cool you off rather than improve the distribution of warm air.

Mounting Location

With the low ceilings common in homes these days, there isn't much leeway for mounting a ceiling fan. The hardware supplied with most fans places their blades about a foot below the ceiling—a measurement you can't reduce much without impairing the fan's operating effectiveness. In many rooms, the blades will be spinning only 7 feet above the floor. That's the bare minimum height for safety.

A fan should not be located where it could sweep curtains and the like up into its blades. When figuring clearances, you also must take into account the headroom under the center of the fan—usually a few inches less than the height of the blades because of projections at the hub. If you use a ceiling-fan kit that includes a light under the fan, you'll lose even more clearance.

Installation

Strictly speaking, a ceiling fan should be installed only by a licensed electrician. But many householders still go the do-it-yourself route; fan makers often promote their products' ease of installation, and their instructions don't read as though they are aimed at professionals.

A fan is easiest to install if it can replace an overhead light fixture, since the wiring already is there. Failing that, you'll have to fish new wires through the walls and ceiling (no job for an amateur) or run the wiring up the wall and across the ceiling inside a special conduit or camouflage the wire in a decorative swag chain.

Don't try to hang a ceiling fan from a standard electrical box; it won't meet the 35-to-50-pound load rating that's required. The fan must be either suspended from a ceiling joist (or crossbrace) or connected to an electrical box specially designed to carry the weight. Check with the dealer about such boxes.

Multispeed ceiling fans usually have a pull chain for turning them on and off and selecting speeds. A more convenient location than on the fan itself is in a wall switch that supplies power to the fan. That's fine if the fan is replacing a light fixture, since the switch wiring already is in place, but things aren't so simple if a fan includes a lighting fixture and you want to control fan and lights independently on single-circuit wiring. Some helpful control devices are available. Ask your dealer (or an electrical supply store) about them.

Some fan makers tell you to attach the blades to the hub after you've hung the motor from the ceiling. It's a nuisance to work overhead, and you may need help, but when you follow this sequence, you do avoid the risk of damaging the blades while jockeying the motor into position. Take note that some expensive fans come with matched, balanced blades: If you damage one, you may have to replace the entire set. Out-of-true blades will be unbalanced, often causing a fan to wobble—among the most common complaints about ceiling fans.

EIGHT

Saving Energy Dollars with Appliances and Lighting

Nine appliances account for nearly three-fourths of all the energy consumed in U.S. homes: refrigerators; refrigerator/freezers; freezers; dishwashers; clothes washers; clothes dryers; room air conditioners; furnaces; and water heaters. In the late 1970s, the major-appliance industry was under pressure to reveal, model by model and in a way that was clear to consumers, how much energy its products used. The idea was that labeling would stimulate competition on the basis of energy efficiency. Appliance labeling was mandated by Congress as part of the Energy Policy and Conservation Act of 1975. The U.S. Department of Energy and the Federal Trade Commission jointly manage its implementation. Energy-cost labels now are attached to many appliances in the marketplace, and they are required on the nine major appliances noted above.

Samples of a refrigerator/freezer label and a room air conditioner label are shown here, with energy costs adjusted to reflect 1990 averages.

The large number in the center of the label is the estimated annual cost ($145) of the energy required to operate the appliance, based on the 1990 national average electricity rate of 7.9 cents per kilowatt-hour. The bar beneath it shows the range of operating costs of competing brands and models of similar sizes and features. The label does not, however, indicate the brand and model number of those other models. You can obtain that type of information by sending $1 to the Association of Home Appliance Manufacturers (AHAM), 20 North Wacker Driver, Chicago, Illinois 60606; request AHAM's latest directory of certified refrigerator/freezers.

Energy-cost labels for dishwashers and clothes washers also include a chart that indicates the cost of running the appliance with a gas or an electric water heater. A representative of your utility company can tell you the energy rate for your area, or you may be able to check it for yourself from your utility bill.

Climate-control appliances, such as room air conditioners, are labeled

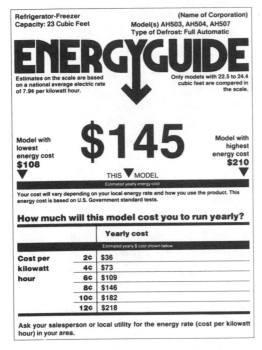

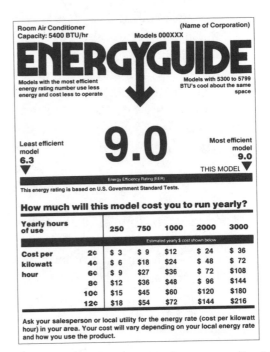

Figure 8.1 Refrigerator/freezer energy guide   **Figure 8.2.** Room air conditioner energy guide

with an energy efficiency rating (EER) or a seasonal energy efficiency rating (SEER). The rating is printed in large numbers in the center of the label. The higher the number, the more efficient the appliance. A range is also supplied—printed just below the energy efficiency rating—for competing room air conditioners of similar cooling capacity. As with refrigerator/freezers, however, the label does not indicate the brands and model numbers of those other models. This information is also obtainable from AHAM (see address above). The organization's directory of certified room air conditioners costs $1.

The U.S. Department of Energy describes a simple way to figure which appliance is more economical. Let's assume that two refrigerators are being compared, each having the same capacity and features but with different purchase prices and operating costs.

Use the following three steps to see how long it will take for a more expensive, more energy-efficient model to pay for itself. First, subtract the price of the less-expensive model from the price of the more-expensive model. Second, subtract the lower yearly operating cost from the higher yearly operating cost. Third, divide the difference in initial purchase price by the difference in yearly operating cost. The answer will tell you how many years it will take for the lower operating cost of the more energy-efficient model to offset its higher initial purchase price.

The Department of Energy estimates the average life expectancy of water heaters, dishwashers, and clothes washers at 11 years; room air conditioners, 13 years; freezers, 19 years; and furnaces, 20 years. Savings achieved by purchasing an energy-efficient appliance can mount up over the life of the product—especially if fuel prices continue to rise.

REFRIGERATOR/FREEZERS

Some three-fourths of all refrigerators sold have top-mounted freezers. They generally are the least expensive models to buy, and they tend to be more energy-efficient than other types. But the top-mounted freezer tends to be on the small side, and everything in the refrigerator section tends to be just below eye level—a liability that not only is inconvenient but also requires you to keep the refrigerator door open for longer periods than would be necessary in eye-level models.

In bottom-freezer models, the refrigerator shelves are at eye level. These models have a higher list price than those with top freezers, but they tend to have more freezer room. Of course, you must bend down to get at the freezer—an inconvenience if you use your freezer a great deal.

Side-by-side refrigerator/freezers allow you to store more foods at eye level in both compartments. There's more freezer space than in a top-freezer model, as well as multiple freezer shelves arranged for orderly storage and easy access. The narrow compartments also make it easier to find stray items. Side-by-sides cost more to buy and run, however, than do units with top-mounted freezers of similar capacity.

Features

A refrigerator/freezer is the most frequently used appliance in your household; it should have a number of basic conveniences:

■ Make sure the controls are clear and legible—especially the energy-saver switch (for controlling a heater that warms door surfaces to prevent moisture condensation during hot, humid weather).

■ For maximum flexible storage, the more movable shelves the unit has, the better. Removable door shelves are easier to keep clean than the kind that form part of the door liner. On some refrigerators, the door shelves are wide enough to accommodate gallon containers or six-packs.

■ Glass shelves in the refrigerator compartment are easier to live with than wire shelves. Spill some liquid or food on a glass shelf and it stays there. Spill it on a wire shelf and it trickles down onto anything below it. You can also wipe up a spill more easily from glass.

■ Meat-keepers and vegetable crispers should glide in and out easily for easy storage and cleaning. The butter compartment should have a tray; its lid should be transparent and stay up by itself. Eggs are safer stored in a tray or bin than in an egg nest on the door.

■ All parts of the unit should be well lighted, and the light bulbs should be properly shielded but easy to reach for changing.

■ The inner and outer doors should be conveniently easy to open and close, and handles should provide enough knuckle space to prevent finger rings from scratching the door surface. (Be sure to order your new refrigerator with the door mounted to swing open in the appropriate direction for your particular kitchen arrangement.)

■ A condenser coil (the part of the refrigerator that disposes of the heat removed from inside the food storage compartments) and drip pan that are behind the front vent at the unit's bottom usually are easy to reach, but they need cleaning about every three months, according to the manufacturers. Dusty coils make the refrigerator run longer, costing you more in electricity and shortening the life of the appliance. Back-mounted coils collect less dust, but you have to move the refrigerator to reach them.

Inside Temperature

When you buy a new refrigerator/freezer, equip yourself with a pair of refrigerator/freezer thermometers as well. They will help you set the appliance for the best balance between performance and economy. A freezer should be set to hold foods at 0°F. If it's colder than that, energy costs rise; if it's a few degrees warmer, then the storage life of your food will be reduced. The refrigerator section, for its part, should be set to store food at 37°F. That's a reasonable compromise between near-freezing—at which meats and most fruits and vegetables do best—and the 40° or so that's appropriate for other foods. Because a change in control setting for one compartment may affect the temperature of the other one, it may take some trial-and-error adjustment to achieve that 0°/37° balance between the two.

You may have to repeat this balancing act several times a year, depending on your refrigerator's built-in ability to compensate for seasonal temperature changes. A refrigerator may have to cope with swings in kitchen temperature of as much as 20°, which could affect the unit's interior temperatures by as much as 5°. Such a change calls for adjustment of the controls.

Keeping the freezer compartment as full as possible reduces the influx of warm air each time the door is opened.

Using a refrigerator/freezer thermometer. Liquid-in-glass thermometers show temperature changes by the movement of a colored liquid in a glass tube; in dial-type thermometers, the expansion and contraction of a metallic coil moves a pointer around a dial to indicate temperature level. Whatever the design, most models are marked in both Fahrenheit and Celsius scales, usually in 2° or 5° increments. And most are designed to stand on or hang from a convenient shelf.

If you have a single-control refrigerator, put the thermometer near the center of the main space and away from any bulky foods. Let the temperature stabilize for several hours during a period when you know the door will not be opened—overnight, for example. Then, take a reading as soon as possible after you open the door. The faster you take the reading, the less likely that warm air will affect it. Keep adjusting the control until you consistently get a reading of 37°F. If the freezer space then isn't reasonably close to 0°F, you may have to readjust until you come close to a consistent 37°/0° combination.

Things may get trickier with a two-control refrigerator. Although the controls are separate, they don't work totally independently, and a change of setting for one compartment may affect the temperature of the other. Getting a balance of 37° and 0° at the centers of the two compartments is apt to take more successive adjustments than with one-control models.

For Ratings of refrigerators with top-mounted freezers, see Appendix B.

FREEZERS

As with refrigerators, the type of freezer you buy has some effect on performance and energy efficiency. Consumers Union's tests suggest that chest freezers (models with a lid on top) are slightly preferable to uprights (models that open at the front).

In a test comparing the two types, chest freezers proved to be more energy-efficient than uprights. (Chests usually are less expensive to buy, too.) The design of a chest freezer helps explain why it tends to have slightly lower energy costs. When you open its lid, cold air tends to stay put. But when you open the door of an upright, the colder air at the bottom spills out and warmer air moves in to take its place.

"No-frost" freezers, which have become increasingly popular over the years, are more expensive to operate than the manual-defrost variety, adding about $30 worth of electricity each year for the convenience of automatic defrosting.

Most freezers use hot tubes inside their walls to warm the cabinet surface slightly and therefore prevent condensation from developing in warm, humid weather. Some models have electric heating elements that can be switched on to reduce condensation—and off again to cut the operating cost when the weather turns cooler and the humidity is low.

As with refrigerators, the internal temperature of a freezer may change with the seasons. In spring and fall, it's best to reset the controls to maintain 0°F by using a refrigerator/freezer thermometer.

Convenience and Maintenance

Freezers with a plastic or porcelain-on-steel interior resist scratching, chipping, and rusting better than those with painted-steel insides.

Most freezers have an interior light. The bulb should be designed so that food packages don't bump into it. Some models have an exterior light that indicates that the power is on, to provide a warning of an accidental unplugging or an overlooked blown fuse or tripped circuit breaker.

Space organizers are minimal in chest freezers—usually a wire basket and maybe a divider. Most uprights have three fixed shelves in the main compartment.

A typical upright has five solid shelves in the door. The metal retainers that keep cans and packages on the shelves generally are fixed.

Defrosting

The thicker the layer of ice that accumulates inside a freezer, the less efficient the heat transfer between the cooling coils and the food. Lowered efficiency means longer running time for the compressor and thus greater electricity consumption. But a freezer isn't opened nearly as often as a refrigerator, so it may not pay to invest in the extra initial cost (and the extra running cost) of a freezer with a self-defrosting feature.

You probably will have to defrost a non–self-defrosting upright freezer twice a year, the frequency depending on the humidity level and on how often you open the door. With a chest freezer, however, you may be able to go 12 to 18 months between defrostings, especially if you occasionally scrape off the frost that builds up around the rim.

Any freezer that requires defrosting should have a drain to draw off the water coming from the melted ice, plus a hose attached to the drain.

Managing a Freezer

Nutritionally, meat, fish, and poultry are the same frozen or fresh. Fruits and vegetables can lose vitamins if they're not handled properly before freezing.

The wrapping for the food needs to be vaporproof, such as Saran Wrap. You also can use aluminum foil or plastic containers with snap-on lids. Don't rely on waxed paper, butcher's paper, regular polyethylene plastic wraps, or even cardboard ice cream cartons. Rewrap all supermarket-packaged meat. Before you seal the package, try to expel as much air as possible. Freezer tape, rubber bands, twist ties, or even string can seal the wrapping.

DISHWASHERS

If you don't rinse the dishes before you load your dishwasher—and you need not—a dishwasher actually uses no more water than hand-washing dishes in a sink full of water—and it also uses less water than washing dishes under a running faucet. The machines themselves also require only a small amount of electricity; in Consumers Union's 1990 tests, dishwashers consumed between 0.6 and 1.1 kilowatt-hours of electricity when supplied with 140°F water. That works out to between 4½ and 8½ cents of electricity, at average power rates at the time. No-heat air drying, which works on evaporation and heat retained from the wash, produces reasonably dry dishes, provided you can wait a few hours before using them or putting them away. This approach to drying saves a penny or two a load.

The bulk of a dishwasher's operating cost is expended by heating the water to feed it. An electric hot-water heater will use 18 cents worth of electricity to provide the 11 gallons of 140°F water that the typical dishwasher uses for one load; assuming you run the dishwasher once a day, the total comes to about $90 a year. The comparable hot-water cost for a gas- or oil-fired heater will be about 6 cents a load, or $45 a year. A number of dishwasher models can help you cut your utility bills slightly by using water supplied at 120°F instead of the usual 140°. The dishwashers then will boost the water temperature as needed, either at the push of a button or automatically. Setting your water heater's thermostat to 120°F should cut those totals by a few dollars a year.

Shopping for a Dishwasher

Most dishwashers offer some variations on the basic wash-rinse-and-dry. A dishwasher's normal or regular cycle typically includes two washes interspersed with two or three rinses. A heavy cycle can entail longer wash periods, a third wash, hotter water, or all of the above. A light cycle usually includes just one wash. These are the basic cycles—and probably all that are necessary. Nevertheless, additional washing and drying options abound.

The common rinse-and-hold option can be useful for small families. Instead of stacking dirty dishes in the sink or dishwasher, you can gradually accumulate a full load, rinsing the dishes as you go. But don't expect a machine that offers a pots-and-pans cycle to do work that requires abrasive cleaners and elbow grease. And think twice before subjecting good crystal or china—especially sets with gold trim—to a dishwasher's china/crystal setting. The harsh detergents and possible jostling could etch or otherwise damage fine glassware and dishes.

Safety

If you should open a dishwasher in midcycle—to add a forgotten plate, perhaps—your chances of getting splashed are nil. All models have a safety interlock to cut power when the door is opened, and most have latches that prevent you from opening the door too quickly. All models also have a float switch, which senses accidental overfilling and automatically turns off the power.

Many dishwasher accidents involve people cutting themselves, usually on knives or forks as they reach over a flatware basket into the machine's dish rack. To avoid this, it's always a good idea to load flatware with its points down. Be aware, too, that most machines have an exposed heating element under the lower rack which can inflict a serious burn. Make sure it has cooled before you reach into the bottom of the tub to clean a filter or retrieve an item that has dropped.

Keep children away while the dishwasher is running: Door vents, often at a toddler's eye level, can emit steam. Some electronic models have a hidden touchpad that locks the controls to discourage kids from playing with them. That's a very worthwhile feature.

For Ratings of dishwashers, see Appendix B.

CLOTHES WASHERS

The main energy cost in a clothes washer lies in the hot water it uses, so operating it prudently offers an opportunity for saving.

Because the amount of hot water a clothes washer uses is a meaningful gauge of its energy efficiency, using as little hot water as possible saves energy. Most machines give you several ways to combine temperatures for wash and rinse water—typically, hot/cold, hot/warm, cold/cold, warm/warm, and warm/cold. If you have a load of unusually dirty clothes, use a hot wash (if the fabrics can tolerate it); otherwise, a warm or cold wash may do the job. And always use a cold rinse: Warm water doesn't rinse any better, and it may increase wrinkling of permanent-press fabrics.

In addition, virtually any modern washing machine will let you control the amount of water you use. If you have a large load of laundry to do, set the water control for maximum fill; if you have a smaller load, set it for proportionately less. In most models with a large tub capacity, the difference between the maximum and minimum fill ranges from 20 to 30 gallons. For best economy, remember that washing a few large loads is more advantageous than washing the same amount of clothes in many small loads.

Although top-loading washers dominate the marketplace, some front-loaders are still available. While these use less water per unit volume of laundry, and less hot water, than top-loaders, they also are expensive, hold relatively little laundry, and may create more repair headaches than a top-loader.

If you machine-dry your laundry, you may find a hidden energy savings with certain clothes washers. Those that extract water particularly efficiently will yield clean loads that shorten the running time of a dryer—and thus reduce operating cost.

As for washing ability, if you use a good detergent and avoid overloading the tub, any washer should clean ordinary laundry effectively.

What's Available

Clothes washers range in price from under $300 to more than $800, the price depending on some of the variations and options available.

Certain special models of washers are designed for particular installations—compact ones that stack with a dryer or have a built-in dryer on top, or rolling ones that hook up to the sink, for example. But most washers (and their companion dryers) are "full-sized" models—27 inches wide. You usually can choose between "large" and "extra-large" tub capacities. Large models typically run $350 to $450; extra-large models are $400 to $600. In general, washing larger loads uses water, energy, and detergent more efficiently than washing smaller loads.

Features and Conveniences

As with most products, the more you pay, the more frills you get. Here's a rundown:

Extra temperatures. All you need are the three basic wash/rinse choices—hot/cold, warm/cold, and cold/cold. Fancier models usually offer hot/warm and warm/warm cycles as well. Warm-rinse cycles are unnecessary—wasteful to buy and even more wasteful to use.

Extra cycles. The basic offerings usually are regular, permanent press, and delicate. Sometimes the choice of cycle automatically determines the speed and the water temperature, but often you set all three yourself. Fancier models add a soak or a prewash stage before the wash or an extra rinse at the end—something you can do with any washing machine just by manipulating the dial.

Extra speeds. A second, slower motor speed allows certain laundry to be treated more gently. The slower speed can be used for the agitation part of the cycle, the spin part of the cycle, or both. Normal-speed agitation with normal spin is fine for most clothes. A slow agitation with a slow spin is useful for washing delicates. These two choices are all that most people need. If you line-dry your clothes, you may appreciate a third choice—normal agitation/slow spin, which leaves permanent-press items less wrinkled than a normal spin does.

Water levels. Since adjusting the water level to the load saves water, energy, and detergent, a choice of at least three fill levels is useful. Fancier models offer four, or have a control that is continuously adjustable.

Finishes. Traditionally, the toughest finish for a washer's top was porcelain, with baked enamel running a cheap and distant second. Newer plastic-based finishes, softer than porcelain but harder than enamel, are now common. They often have a trademarked name, such as Dura-Finish or Endura-Guard.

Electronic controls. You'll find these only on top-of-the-line models. They make a simple task unnecessarily complex. An electronic touchpad may be slightly easier to manipulate than the knobs most machines use for controlling temperature, water level, and the like. But electronic cycle selectors often tell you less about cycle choices and the duration of the cycle than the old familiar dial.

Handling unbalanced loads. A wad of wet towels or blue jeans that throws off the balance of the spinning tub can make the quietest washer clank, bang, and sometimes take a walk. Many models still can't cope well with unbalanced loads, as the Ratings show. Sometimes a problem can be minimized by making sure the washer is level. Many models now use a clever design that links the rear legs together to make leveling easier. A wood floor with too much flex can exacerbate vibration and noise, especially if the frequency of the washer's shaking is in tune with the natural resonance of the floor, much as a guitar body amplifies a plucked string. Slipping a thick plywood panel under the washer or stiffening the floor joists with blocking or extra bridging may help.

For Ratings of clothes washers, see Appendix B.

CLOTHES DRYERS

Nearly all the dryers available in the stores let you set temperature ranges that correspond to the needs of different fabrics; most also have automatic-drying cycles that turn off the heat when the clothes are dry.

The oldest, simplest approach to automatic drying—and the one that most moderately priced dryers employ today—uses a thermostat to control the heat and the timer. As the air leaving the drum becomes warmer and the dryer's heat drives moisture from the clothes, the thermostat samples the air temperature. When the temperature climbs to a certain point, the thermostat shuts off the heat and advances the timer. Then the heat goes on again, and the timer stops. This on/off sequence continues until the timer has finished its cycle.

Newer, more expensive dryer designs sense temperature and moisture with better precision, using sensors inside the drum that touch the clothes and so detect dryness directly.

Drying Techniques

Cottons can take a fair amount of heat without suffering, but permanent-press items need to be warmed enough to relax wrinkles, and then cooled down before new wrinkles are allowed to set.

The permanent-press cycle on a typical clothes dryer runs at a cooler "medium" heat setting than a "regular" or cotton cycle. Lingerie and other "delicates" need very little heat—probably not much more than 110°F. Rubber and plastic items shouldn't be subjected to any heat all all, just a tumble in the fluff or air-dry cycle.

Some dryers run quite a bit hotter in the permanent-press cycle than others, but that shouldn't cause a problem. The temperature of the clothes at the end of the cool-down phase at the completion of the dryer cycle seems to make more difference to the final appearance of permanent-press items than the temperature reached during the drying sequence.

Permanent-press items left in a heap in the dryer will wrinkle even if the dryer leaves them at room temperature. To minimize wrinkling, some machines have an extra tumbling-time feature called "wrinkle guard," "press guard," or something similar. It tosses the clothes for 15 minutes to a couple of hours after the heat shuts off. Some machines toss the load constantly, some intermittently. Either way, the added tumbling can help.

Features

Temperature control. The presence of electronic touchpads may let you control temperatures and drying times precisely, but they add to a dryer's cost and can be expensive to service. Conventional knobs work just as well.

Filters. A lint filter is best located in the drum opening, where it acts as a reminder for cleaning each time the dryer is used. As far as drum size is concerned, the bigger the better. And an interior light is a nice convenience.

Doors. Most dryer doors open to the side (hinged at the right, the assumption being that a companion washing machine is on the left). A side-opening door lets you put a laundry basket up close to the dryer's opening. A door that opens down creates a handy shelf, but it also can force you to stretch to reach into the drum.

Energy efficiency. Clothes can't tell if they're being dried with electricity or gas, but the difference shows up on your utility bill.

Gas dryers can cost some $35 to $65 more than electrics initially, but you would quickly earn that difference back in lower operating costs, assuming that the appropriate service lines were already in place. Using 1990 national average costs (7.9 cents per kilowatt-hour for electricity, 56 cents per 100 cubic feet for natural gas), a family that dried eight loads of laundry per week would spend $97 each year running an electric dryer and $41 operating a gas dryer.

Here's a rule of thumb for those whose rates differ from the national average: A gas dryer will cost as much to run as an electric dryer when the cost per 100 cubic feet is about 20 times the cost per kilowatt-hour of electricity. When the cost per 100 cubic feet is less than 25 times the cost per kwh, gas is cheaper; when the cost is more than 25 times the cost per kwh, electricity is cheaper.

To save energy, set a dryer's automatic control for as low a dryness setting as will provide proper drying; with a timer control, be careful to avoid over-setting. Both of these methods will require some experimentation on your part. Another way to cut energy costs is to use a clothesline or drying rack to dry such items as bath mats, which hold a considerable amount of moisture. In fact, for maximum energy saving, use a clothesline or drying rack for as much of your wash as is convenient.

Venting

A typical 12-pound load of laundry weighs about 20 pounds as it comes from the washing machine—meaning that a dryer has to dispose of about a gallon of water with each such load. The resulting moisture in the exhaust air virtually demands that it be vented outdoors. Although venting a dryer indoors (with an appropriate filter for removing lint from the air) may seem an attractive option because the dryer's heat could act as a house-heating supplement, mildew and other problems can occur with so much moisture being pumped into the air over a short time period.

Flexible plastic duct pipe—inexpensive and easy to install—might seem just the thing for venting a dryer that must be located some distance from an exterior wall. It isn't. That flexible pipe, which resembles a plastic-covered Slinky toy, can create a significant fire hazard, according to several dryer manufacturers.

The problem is that plastic vent pipe lends itself to long runs and snaking droops that constrict the flow of air. That causes a buildup of lint in the pipe, makes the dryer run hotter, and, in time, increases the dryer's running time (and its energy costs).

A safer choice, particularly with long runs of ducting, is galvanized-steel duct pipe. The rigid pieces of pipe fit into one another snugly. Galvanized pipe does cost more, but it won't kink or collapse the way a flexible plastic pipe will. Next best is flexible metal ducting, sold at large hardware and plumbing-supply shops.

For Ratings of clothes dryers, see Appendix B.

KEEPING WARM IN BED

Until several years ago, conventional blankets were the most popular kind of bed covering. As fuel prices rose, however, and householders set their thermostats to lower nighttime temperatures, comforters reemerged as the bed covering of choice, and their sales began to climb, quickly equaling those of conventional blankets and electric blankets combined.

Pound for pound, a comforter provides a lot more warmth than a pile of blankets, and it's also more convenient. Not only does it cover the sleeper at night, it also covers the bed, in lieu of a bedspread, during the day. Making up the bed in the morning takes just seconds: Fluff and smooth the pillows and the comforter, and the job is done.

Polyester-filled comforters outsell the down variety: They're less expensive, on average, and they look better. Manufacturers also coordinate polyester-filled comforters with sheets, package them with dust ruffles and pillow shams, and match them to drapes and even tablecloths.

Electric blankets and electric mattress pads have a small but loyal following; about 60 percent of sales are replacements. But electric bedding has an image problem: A lot of people have never liked the idea of sleeping on or under live wires, fearing electric shock—one reason manufacturers took to substituting the word *automatic* for *electric*. Now there are additional concerns about the biological effects of low-frequency magnetic fields produced by electric blankets. (See the discussion of this later in the chapter.)

COMFORTERS

A typical polyester-filled comforter will provide as much warmth as a typical down comforter. The warmest comforter—filled, perhaps, with expensive goose down—may equal the warmth of more than four woolen blankets. But many goose-down models are not at all warm.

The content of the filling material alone is no measure of warmth. Other important factors are the *amount* of filling and the construction of the comforter. You can judge the warmth of a comforter in the store by comparing thickness (called *loft*); a thick comforter usually traps more air than a thin one, so it insulates better. Trapped air is what retains warmth.

A warmer comforter generally is plumper than a less-warm one. Differences can be dramatic among down comforters, where loft may range from 3½ inches in a warm model down to ½ inch in a comforter that provides very little warmth. The range of differences among polyester-filled comforters is likely to be less, but it's still significant.

When you are comparing comforters in a store, compress a center section of each between your hands. You'll feel the difference between a substantial filler and a skimpy one. Don't measure the filling near the edges of the comforter. In cheaper comforters (particularly those filled with polyester), the edges frequently are overstuffed.

A lamb's wool or silk-filled comforter often is thinner than a down comforter of equal warmth. But it's heavier and might weigh a pound more.

Some comforter labels refer to *fill power*—an industry measurement of down quality. An ounce of down is placed in a cylinder; the fluffier the down, the more space it fills. Fill power may range from 300 cubic inches per ounce to as much as 650 to 700 cubic inches per ounce. The measurement is useful to manufacturers—the better the fill power, the less down is needed to fill a comforter. But if you don't know the weight of the filler (and it's rarely noted on a comforter's label), fill power doesn't tell you much about warmth.

Other Aspects of Comforter Warmth

The warmest down comforter has fabric baffles—like small interior walls—connecting the comforter's top cover to its bottom cover. The baffles allow the down to achieve its full potential loft. The presence of baffle construction usually is noted on a comforter's label.

A warm down comforter will be sewn in a tufted fashion. In down-comforter marketing, tufting is often called *karostep* stitching (*karo,* in German, means diamond). According to the manufacturers' assertions, the advantage to tufting is that it lets you move the down filling to where you need it most. More important, it gives the filling more room to fluff up for maximum warmth.

Polyester comforters usually don't have tufted quilting. But if you find one that does, its warmth could be superior to a comparable model without quilting.

Most comforters—polyester and down—have sewn-through quilting (i.e., the top cover is attached to the bottom cover), often in a channel or box-stitched design. The claim with channel or box stitching is that it holds the filler in place, so you'll never find any cold spots. But a more important characteristic—and a negative one—of such designs is the compression of the filling near the stitches. That limits the loft of the filling and, hence, the amount of air it can trap.

The way a comforter conforms to your body has some effect on how warm you feel underneath. A down-filled model tends to drape better than the polyester variety.

Other Features

Covering. The manufacturer may refer to the cotton covering on a down-filled comforter as "down-proof." That means simply that the fabric is tightly woven (230 or more threads per square inch) to prevent down and feathers from escaping.

Color. Dark-colored fabrics have a tendency to shed some of their color. This doesn't mean that a noticeable amount of color will rub off on nightgowns and pajamas. It does mean, however, that a dark color—wine, navy blue, black—is not likely to remain the same shade after several washings or dry cleanings.

Comforter care. Some manufacturers offer you a choice between dry cleaning and laundering. Even among down-filled comforters that supposedly are washable, significant shrinkage can occur—perhaps 6 inches in length or width.

Washing a comforter is a lot cheaper than having one dry cleaned, of course, but you'll probably have to spend half a day at the neighborhood laundromat to do the job, since you'll need the extra size of a commercial machine. Before you commit to washing, make sure the laundromat has a dryer that runs at a reasonably cool temperature. Many commercial dryers run too hot to dry a comforter safely.

Given the inconvenience of laundering—and the potential for shrinkage with a down-filled comforter—you may be willing to pay the price for professional dry cleaning.

You can reduce the need to clean a comforter if you use a *duvet cover*— essentially a pillowcase for a comforter. Then you can launder the cover regularly and simply air out the comforter as needed.

Duvet covers usually are expensive. As comforters have become more popular, the covers have become more available, but not cheaper. You should be able to find duvet covers in linen stores, mail-order catalogs, and even department stores. They range in price from about $50 to well over $100. Matching pillowcases are also available.

Recommendations

A comforter is the most convenient way to keep warm in bed. There are no wires to worry about and no dos or don'ts to be concerned about.

Don't be tempted to buy more comforter than you need. Advertisers of down sleeping bags often note that each inch of down equals about 27°F in thermal protection. If a down comforter with 1-inch loft keeps you comfortable in a bedroom where the temperature is about 72°F, one with a 2-inch loft will keep you warm in a room with a temperature as low as about 45°F.

The loft-to-warmth rule isn't as clear-cut with a polyester comforter as it is with down, but a thicker model will still be warmer than a relatively thin one.

A silk- or lamb's-wool-filled comforter isn't likely to be an improvement over other types. Either type is likely to weigh more than a typical down or polyester comforter and may not drape as well as a down model.

Down and Polyester

A comforter labeled *down* is required by federal law to contain at least 80 percent down. The other 20 percent can be made up of small feathers, tufts of imperfect down or feathers, or even small amounts of dust, the consequence of imperfect processing. A comforter labeled *all down* or *100 percent down* is required to be exactly that.

A *feather-and-down* comforter can be any combination of feathers and down. But if the manufacturer specifies a certain percentage of each, the comforter's filling must live up to the claim.

Down is soft, fluffy, and light and it does an excellent job of trapping air. *Feathers,* in contrast, are stiff, flat, and heavy—far less effective in trapping air. In a comforter with a substantial feather content, you may be able to feel the bits of rigid—sometimes sharp—feather quills through the comforter cover.

Down from the eider—a wild duck native to the arctic and other northern climates—is considered to be the best. If you have a few thousand dollars to spend, you can buy an eiderdown comforter. The astonishing price of eiderdown is less a reflection of its quality than of its scarcity. The eider is endangered, and thus protected, so the down has to be collected from the ducks' nests—no easy task.

Goose down is far less expensive. And while it has a certain mystique that implies its vast superiority, it's only slightly warmer, ounce for ounce, than duck down. If a comforter is labeled *goose down,* the law requires that it be 90 percent goose down. A comforter labeled *down* may contain goose down, duck down, or feathers from domestic fowl such as chickens or turkeys.

The United States has no regulations governing the labeling of country of origin of the filling material. Labels that promote European goose down do so in order to justify a high price. There's no quality difference among European down, domestic down, or Asian down (the most common kind nowadays).

Some comforter labels boast *all-white down.* The only advantage of white down is that it's less likely to show through the covering material than brown or gray down. The color of the filler does not affect its warmth.

Polyester labeling is less complicated than that for down. Despite the number of brand names—such as Kodel, Trevira, Fortrel, Hollofil II, Kodofill, Quallofil—most polyester fillings are just plain polyester. Some consist of fibers with a hollow core that is claimed to enhance the material's warmth-retaining properties.

ELECTRIC BLANKETS AND MATTRESS PADS

The even distribution of heat in an electric blanket is created by wires that run in parallel lines from the top to the bottom. In most electric blankets, the heat is zoned: There are more wires in the lower half of the blanket than in the upper half (feet and legs tend to become cooler faster than the upper body).

In many electric blankets and mattress pads, thermoswitches connecting the wiring prevent overheating. As a substitute for thermoswitching, a blanket may use wiring joined together by electrically conductive plastic. As the plastic heats up, it becomes resistant to electric current and, in effect, turns down the heat in hot spots.

Electric blankets and pads don't become very warm to the touch. Their function is to slow down the escape of the sleeper's body heat rather than to provide actual warmth, the way a heating pad does. Most blankets reach their maximum temperature (between 80° and 89°F) in about 15 minutes. The maximum temperature in mattress pads ranges from about 72° to 79°F. (An electric mattress pad doesn't have to heat up as much as an electric blanket because the heat it generates, as well as the sleeper's body heat, is trapped under covers.)

You would be too warm if you slept through the night with an electric blanket or pad operating at maximum temperature. But if you like to pre-warm your bed, a model with good heating performance will do the job fastest.

Controls

The controls on most electric blankets and pads make periodic clicking sounds as they cycle on and off. The noise isn't loud enough to be annoying unless you are a light sleeper. In that case, you might prefer a blanket or mattress pad with a silent control.

Controls contain an on/off switch and temperature settings ranging from low to high; they usually are backlit for visibility in the dark. You may even find a deluxe model with bright and dim settings. However, dials that are easy to read from the top may be hard to read from the side, and vice versa.

It's a good idea to inspect the controls of any blanket or pad before you buy. Make sure the markings are large enough to read easily. And try to operate the control with one hand. Some blankets come with control hangers, so you can hang the control over a bed's side rail or headboard, or hook it onto a nightstand's drawer pull.

Fabrics

Most electric blankets are a blend of polyester and acrylic. Polyester is a sturdy fiber that lends strength to the fabric; acrylic provides softness. Most blankets come in shades of blue, pink, gold, and white.

Mattress pads come in all-polyester, polyester and acrylic, and polyester and cotton.

Daily handling and regular laundering can abrade a blanket's fabric, causing pilling that makes it look old before its time. Pilling is more likely to occur in blankets with the highest percentage of polyester and the lowest percentage of acrylic.

Caring for Electric Blankets and Pads

Electric blankets and mattress pads require a certain amount of respect, and they should be used with care.

■ Don't cover an operating electric blanket with a pillow or other bedding.

■ Don't sit down or lie down on a working blanket. (Don't let a dog or cat do it, either.)

■ Don't fold or bunch up the blanket when it's operating.

■ When you make the bed, make sure none of the wired parts are tucked under the mattress.

■ Launder blankets and mattress pads rather than dry-cleaning them. Cleaning fluids can damage the wiring.

■ Make sure the plug joining the control to the blanket or pad is fully inserted or firmly connected. In a number of models, the pins in the connector are recessed beneath a shield. With that design, fingers and fibers are less likely to come into contact with a live pin, so it averts the risk of an electric shock.

Before you put away an electric blanket or mattress pad for the summer, inspect its wires, plugs, and controls to make sure everything is in good shape. If something is amiss, the manufacturer's tag, sewn to the blanket or pad, tells where to send the appliance for repairs. When folding a blanket or pad for storage, avoid kinking the wires.

A New Safety Concern: Electric and Magnetic Fields

A debate is ongoing about the safety of the electric and magnetic fields produced by power lines, household wiring, and appliances—especially electric blankets and related products. They have been singled out because the user's body stays very close to these products' wiring for significant periods of time (all night, every night, for months at a time every year).

Some scientists contend that electric and magnetic fields affect cells, alter moods and biological rhythms, and promote cancer. According to them, the evidence is alarming enough to justify rerouting transmission lines and redesigning appliances. Yet others who have looked at the same evidence contend that the fields are harmless.

The invisible fields at the focus of this dispute are created by the alternating current that powers the nation. Alternating current changes direction 120 times, or 60 full cycles, each second—a frequency referred to as 60 hertz (Hz). The resulting electric and magnetic fields pulsate at the same rate and are known as 60-Hz fields.

These low-frequency fields transmit very little energy compared with other forms of electromagnetic energy. Unlike X-rays, for example, which have frequencies trillions of times higher, such fields can't break chemical bonds and cause mutations that might lead to cancer and birth defects. Unlike intense microwaves, they don't heat tissue.

Nevertheless, in 1987, prompted by growing concerns about 60-Hz fields, Congress's Office of Technology Assessment authorized an in-depth review of research on 60-Hz fields. The study was carried out by Carnegie-Mellon University's Department of Engineering and Public Policy and published in May 1988. Some of the study's principal conclusions are:

"The emerging evidence no longer allows one to categorically assert that there are no risks" from 60-Hz fields. "But it does not provide a basis for asserting that there is a significant risk."

Numerous experiments have looked for effects from 60-Hz fields without finding any. But a "growing number of positive findings have now clearly demonstrated" that low-frequency fields "can produce substantial changes at the cellular level."

Physiological changes in cells or organisms occur at field intensities to which people commonly are exposed. But the effects are not necessarily harmful, and many are reversible when the fields are shut off. Still unknown is whether or not the changes observed have any clinical significance. Key uncertainties still surround 60-Hz fields and their effects on health, however. For example, it's still not known how these fields interact with cells, although it is believed that effects occur at the cell membrane. It's not even clear that weaker fields are safer than stronger ones: A number of studies found that at low intensities, certain effects occurred that failed to take place at higher intensities.

It may be years before any clear answers emerge from other studies that are in progress. But what is clear now is that neither pregnant women nor children should wait for certainty. Fetuses are extremely sensitive to many environmental agents, and it's prudent to assume that they may be sensitive to 60-Hz fields as well.

Consumers Union's medical consultants have recommended that children and pregnant women should avoid electric blankets and pads and stick to the acceptable alternatives—comforters, for example.

Users of electric blankets who are concerned about potential biological effects can use an electric blanket or mattress pad to warm the bed and then unplug the appliance before going to sleep.

For Ratings of electric blankets and mattress pads, see Appendix B.

LIGHT BULBS

Light bulbs always seem to burn out at the wrong time and in the wrong place. Actually, exceptionally long-lasting bulbs do exist, but people usually don't venture beyond the ordinary hardware-store variety when considering how to fill the typical house's 27 sockets. If they did, they would find a very large variety of bulbs available for home lighting. Choosing bulbs carefully for their intended purpose can pay off, considering that more than 16 percent of the electricity used in homes goes into lighting.

The most widely used type of incandescent bulb is rated to last 750 hours. But you can buy bulbs that last thousands of hours, for years of service—extended-life bulbs, for instance. You can opt for bulbs that put out the same amount of light as regular incandescent ones, but for a fraction of the energy, as fluorescents do. Or you can buy bulbs such as reflectors, which focus light where it's needed so that less light is squandered. Most people buy 60- and 100-watt soft-white bulbs, but a variety of choices is available even within the familiar "A-type" incandescent (see the section entitled "Bulb Varieties"): soft white, standard frosted, or clear glass; a range of wattages; or three wattages in one bulb.

Bulb Efficiency

A light bulb's wattage tells the amount of power it consumes, which corresponds only roughly to the amount of light it delivers. Some bulbs are more efficient than others in converting energy to light.

Light is measured in lumens. *Lumens per watt,* which represents lighting efficiency, is analogous to the miles-per-gallon figure for cars and compares efficiency better than lumens alone. Most incandescent bulbs achieve only about 14 to 18 lumens per watt when new. Other bulb types are more efficient. Some halogen bulbs reach about 20 lumens per watt. Fluorescents squeeze 40 and sometimes 80 or more lumens of light from each watt. Highway sodium-vapor lamps can get 100 or more lumens per watt.

Low-wattage bulbs generally are less efficient than high-wattage bulbs. For example, you need *two* 60-watt bulbs to get the same light that one 100-watt bulb provides—and together the smaller bulbs would be using 20 watts more and 20 percent more power.

Bulb Life

As a light bulb burns, tungsten gradually boils off its filament; in time, the filament snaps and the bulb burns out. By varying filament construction and the combination of inert gases inside a bulb, engineers have been able to produce an incandescent bulb that lasts a long time. Some traffic-light lamps, for instance, are rated for 8,000 hours. Yet the life built into regular light bulbs is only 750 or 1,000 hours.

Long-life bulbs for household use have rated lives of 2,000 hours and up, depending on their construction. Such bulbs typically have a longer filament than regular bulbs have; the longer filament burns a bit cooler and is less efficient at converting the power into light. The result of that lower efficiency is that some long-life bulbs may put out only about 10 lumens per watt. As the lumen numbers on the packages reveal, you can expect something on the order of 10 percent less light from long-life bulbs. You can see that difference quite easily in a side-by-side comparison. Over time, however, as the bulb operates, that difference tends to diminish. As a bulb burns, the tungsten filament evaporates and slowly blackens the inside of the bulb, making it dimmer. With an ordinary incandescent bulb, and its relatively short life, the dimming takes place over a shorter time than with a long-life bulb. Therefore, the long-life variety's brightness tends to "catch up" as a result of its slower-evaporating filament.

If your light bulbs burn out much more quickly than the "average life" listed on the package, they may not be dying a natural death. Excessive vibration—caused by children running around on the floor above, for instance—can significantly shorten bulb life, according to the spokesperson for a major lamp manufacturer. If vibration is a problem, you may want to try "rough-service" bulbs, whose filaments are specially supported to better withstand vibration. Such bulbs can be found in electrical supply stores.

Excessive voltage also can shorten a bulb's life. Most bulbs are rated for 120 volts, and they will burn out prematurely if the voltage coming into your house is higher. If that voltage is 130, say, a 100-watt bulb will last only one-third of its rated life, although it will burn 30 percent more brightly. If your bulbs burn out unusually rapidly, call the power company to see if it can regulate the voltage. Or try bulbs rated for 130 volts, also available in electrical supply stores.

If a bulb burns out quickly, you may simply have bought a bad bulb. The package label's "average life" is a prediction, not a promise. Accordingly, some bulbs burn out far sooner than they should, while others glow on and on.

One trick for extending bulb life is to use a switch with a built-in dimmer (either two-position or continuous). Dimmers make the most sense in rooms where you need the light's full intensity only some of the time. But if you can get by on less light all the time, use a lower-wattage bulb.

Power and Bulb Costs

There are two ways to look at how much bulbs cost. One is the cost per year of life—a figure obtained by dividing the price of a bulb by the number of hours it lasts.

But the lion's share of lighting costs—some 80 to 90 percent—is not for the bulbs, but for the electricity used. Energy-saving bulbs save a few watts, but not much money. The five watts of power conserved by an energy-saving bulb translates into less than 40 cents annually, at national average rates for power. Since you pay a premium of up to 25 cents for those bulbs, the net savings is small indeed.

Bulb Varieties

Incandescent A-type bulbs. These are the bread-and-butter of home lighting. Incandescent bulbs, with a tungsten filament surrounded by inert gases, give only 14 to 18 lumens per watt of power consumed. The light is a pleasing color, though. Soft-white bulbs, coated inside with a fine deposit of white silica, have virtually replaced the "standard" bulb, its inside frosted by a fine etching with acid. The soft-white bulb's advantages: less glare and softer shadows. Clear glass bulbs, more akin to Edison's original, still exist, but their light is stark.

Decorative incandescents. Here a great many shapes are possible, in several sorts of glass. Globes are popular for dining areas and bathrooms. Flames and tubes fit special fixtures. Some may be designed to show off the filament.

Tungsten-halogen lights. These are special incandescent bulbs, often produced in strong quartz rather than in glass. Pressurized halogen gas enables the bulb to be efficient (20 lumens per watt) yet small—peanut-size, for some lamps. The light is an intense, bright white. These bulbs often are several hundred watts, so halogen fixtures and lamps often come equipped with a dimmer. The bulbs can last up to 4,000 hours. They must be handled carefully; fingerprints can weaken the surface, which may heat up to more than 500°F and cause the bulb to explode (fixtures ought to shield the bulb). In addition to the many nonstandard sizes available, you can purchase a halogen bulb in the shape of a bulb suitable for an ordinary lamp socket.

Reflector lights. These may be ordinary incandescent bulbs or tungsten-halogen ones. They're mirrored inside and specially shaped for spot- or flood-lighting in track lights or recessed fixtures. A parabolic lens shape *(PAR lamp)* is most typical and comes as a spot or a flood. The elliptical shape *(ER lamp),* designed for recessed "high hats," focuses the light to maximize what gets out of the fixture. Heavy, pressed-glass lamps can be used indoors or out. Blown-glass reflectors generally are for indoor use only.

Fluorescent adapters. These include the "ballast" that fluorescents require to regulate current, allowing you to convert ordinary bulb sockets. A light with an adapter can work in a ceiling fixture or table lamp with a wide shade; a compact fluorescent stick and adapter can work in many home fixtures. The fluorescent-within-a-bulb (inside is a U-shaped fluorescent tube) fits an ordinary incandescent socket. The initial high cost of such specialized bulbs ($10 to $25) is more than offset by their long life—around 10,000 hours—and their greater efficiency. And the adapters are reusable.

Fluorescent tubes. The light in these comes from a special phosphor coating fluorescing inside the tube instead of from an incandescent filament. Fluorescents are remarkably efficient; some give 80 lumens per watt. They use only one-fifth to one-third the electricity of incandescents of comparable brightness, yet they last 10 to 20 times as long. "Cool-white" models give off an unflattering cast, but newer "warm whites" are closer to incandescent in the color they convey. The latest bulbs also are more compact—thinner tubes, only 5 to 7 inches long, for instance. Fluorescent fixtures are especially well suited to area lighting.

Recommendations

Most people want bulbs that last longer. In practical terms, that means buying incandescents that are more expensive and less efficient than regular bulbs, or choosing a type of bulb that differs from the usual incandescent. The real long-life power-misers are to be found among other types of bulbs, especially fluorescents. The energy they save easily can more than offset their higher price.

Regular bulbs are generally least expensive. The energy-saving types save a few watts, but the payoff will be on the order of dimes, not dollars, and you're apt to get a dimmer bulb.

Here's an adapted summary of energy-saving recommendations from the U.S. Department of Energy:

■ Turn off the lights in any room not being used.

■ Concentrate lighting in reading and working areas and where it's needed for safety (stairwells, for example). Reduce lighting in other areas, but avoid very sharp contrasts.

■ To reduce overall lighting in nonworking spaces, remove one bulb out of three in multiple light fixtures and replace it with a burned-out bulb (for safety). Replace other bulbs throughout the house (as they burn out) with bulbs of the next lower wattage.

■ Consider installing solid-state dimmers or high/low switches. They make it easy to reduce lighting intensity in a room and thus save energy.

■ In areas where bright lights are needed, use only one large bulb instead of several small ones.

■ Try 50-watt reflector floodlights in directional lamps (such as pole or spot lamps). These floodlights provide about the same amount of light as the standard 100-watt bulbs, but at half the wattage.

■ You can save on lighting energy through decorating. Remember, light colors for walls, rugs, draperies, and upholstery reflect light and therefore reduce the amount of artificial light required.

■ Have decorative outdoor gas lamps turned off, unless they are essential for safety, or convert them to electricity.

■ Keeping just eight gas lamps burning year round uses as much natural gas as it takes to heat an average-size home for a winter season. By turning off one gas lamp, you might save from $40 to $50 a year in natural-gas costs.

■ Use outdoor lights only when they are needed. One way to make sure they're off during daylight hours is to connect them to a photocell unit or timer that will turn them off automatically.

For Ratings of light bulbs, see Appendix B.

NINE

Contracting Work: When Do-It-Yourself Won't Do

You can manage many energy-saving home improvements—caulking, weatherstripping, even some types of insulation—yourself. And if you're a deft handler of tools and materials, you can do a fair amount of remodeling without outside help. Still, you'll probably want to leave some jobs to professional contractors—installing a wood-burning stove or a solar water heater, for instance. Many local codes *require* that only professionals do wiring and plumbing, or maintain heating and cooling systems.

The home-improvement industry is a chronic source of consumer problems, including shoddy work, jobs left unfinished, and contractors who damage more property than they improve. Nationwide, home repairs consistently have ranked second to automobiles as a source of complaints, according to federal surveys of local consumer-protection agencies.

The problems vary in kind and in size. A homeowner in suburban Port Chester, New York, told Consumers Union that he has had some relatively small but typical problems with contractors over the years. "We hired a sort of general handyman to finish a playroom in the basement," he said. "He hired another guy to install a gas heater, and we had no end of trouble getting the man back to adjust it. The contractor who put up new siding for us did a decent job, but his idea of cleanup was different from ours. He must have left a bushel of stuff behind."

The wise consumer is well advised to be equipped with some basic information before approaching the home-improvement market. If you need to hire a professional, follow the steps we outline in this chapter in selecting one and in negotiating the contract. Also included in this chapter is a form to help you decide on a contractor.

FINDING A CONTRACTOR

The right contractor, of course, is the best possible beginning for a home-improvement project. We suggest you proceed along the following lines.

Identify the job to be done. Well before you even call for an appointment with a prospective contractor, prepare yourself by defining as specifically as possible the work you want done. Write down a rough set of specifications for the job you have in mind. List the work to be done, the kinds of materials you would like to have used, and your budget for the project. If this is your first major piece of home remodeling, take the time to find out what materials are on the market and what they cost. A couple of Saturday afternoons at local home centers or lumberyards should tell you what you need to know. Writing down a rough set of specifications for the job you have in mind and knowing about the cost of materials will help you decide what you *really* want a contractor to do. With a clear picture of the scope of the job, you will be better able to resist a sales pitch for extra work or for the use of more costly materials than you might need. You should, however, expect to alter and refine your initial specifications in the course of discussing the job with contractors.

Ask for bids. Get the names of contractors from friends, neighbors, people at local building supply centers, or your utility company. Then obtain written bids on your job from at least three of them. Don't try to close the deal on a contractor's first visit. Instead, discuss the work you want done, go over your rough specifications, and take note of any modifications or suggestions the contractor makes. As one remodeler said, "This is a time to pick the contractor's brains, not just to shop for price."

On the first visit, the contractor may give you only a very rough cost estimate. That estimate, however, should include a round figure for all the work you want done. If you plan to finance the job, discuss the contractor's financing terms and possible arrangements for payment. As with any major purchase, be sure to shop for the best credit terms. If the contractor gives you a reasonable preliminary price during the first visit, ask for a precise estimate—both for the amount of money the job will cost and the amount of time it will take to do the work. It's also essential to obtain the names of some of the contractor's previous customers. (A useful practice is to request information about jobs similar to the work you have planned.) A check of references is the only effective way to judge a contractor (even though a contractor isn't likely to refer you to a dissatisfied customer).

Visit homes of previous customers. Look over the work and talk with the homeowners. Ask whether the job was started and completed on time, and whether the actual cost exceeded the estimate. (If it did, find out if the overrun resulted from changes requested by the homeowner.) Ask whether the family is pleased with the job and whether the contractor has been willing to fix things that may not have been done properly the first time. Ask, too, whether the contractor spent sufficient time and care in supervising the workers, or whether the contractor's main interest seemed to be in collecting the fee.

Once you've checked the past customers, contact the licensing bureau (if licenses are required where you live) or the consumer protection office in your area. These agencies are not likely to recommend one contractor over another, but you should be able to learn if any complaints have been filed against a contractor, the nature of those complaints, and how they were resolved. While one or two complaints shouldn't alarm you, a string of similar problems should make you reluctant to hire that contractor.

WORKING OUT THE DETAILS

When a contractor passes muster, you're ready to sit down and put your agreement in writing. Go over the job specifications carefully; be sure you understand what the contractor will do, what materials will be used, and when the job will be completed. Review the financing arrangements, the schedule of payments, and any other provisions that should be written into the contract. A written contract is essential for expensive work. It's also important for small home-improvement jobs, because problems can arise easily even on simple or inexpensive projects.

A home-improvement contract actually involves two aspects, one covering the work to be done, the other covering your obligation to pay for the work—whether you're paying cash, borrowing money from a bank or credit union, or having the contractor arrange financing for the job. When you are financing the job, financing or credit terms may or may not be part of the agreement covering the work. If the contractor is handling the financing, the work and your credit obligations may be covered by the same agreement.

Of the two parts of such a contract, what follows concentrates on the one covering the work to be done. The contractor may well have a preprinted form ready for your signature, but you should consider it only a starting point for negotiations. An analysis of several such forms, which was prepared some years ago for Consumers Union by the National Consumer Law Center (NCLC), showed that preprinted forms have one feature in common: From the homeowner's point of view, they leave a lot to be desired.

Most states have laws covering home-improvement contracts in one way or another. Some cities and counties have their own laws, which often are stricter than state laws. It's impossible to catalog all the state and local laws here; for specific information, you should consult a consumer protection agency, local licensing bureau, the state attorney general, or a lawyer.

When it's time to negotiate the contract for the work, be sure you consider the important points described below.

Everything in writing. The contract should include—in writing—everything you've discussed with the contractor. A typical clause might say, "This instrument sets forth the entire contract between the parties and may be modified only by a written instrument executed by both parties." But before you sign a contract with that kind of clause, be sure the contract *does* have everything in black and white. If the contractor sweetens the sales pitch by offering extras that you want, see that they are included in the job specifications; otherwise, you might not get them, or might get them only at an additional price that the contractor then claims was verbally acceptable to you. This clause also protects you if the contractor tries to add on delivery charges or other such costs once work has begun.

Complete job specifications. A former contractor, who went to work for the California board that licenses contractors, told Consumers Union, "I used to spell out everything in my contracts—even the sizes of the nails I'd use. It protected me as well as the homeowner." That much detail might be carrying things too far, but the principle is correct. The contract should be specific. For example: "Install 80 square feet of solar collector on south face of roof, per instructions supplied by collector manufacturer, water-supply and return pipes, ⅜-inch-diameter copper with soldered joints and fittings, to be run through roof and inside house to storage tank in basement; attached drawings specify installation details and exact location." That's much better than specifications that say merely "Install solar water heater." The contract should also specify brand names, colors, grades, styles, and model numbers for appliances and materials. If an architect's or engineer's drawings are required, be sure they are attached to the contract and cited in the specifications, as in the example above.

Starting and completion dates. Both dates are essential because they tell you when the work will take place—and, in some cases, they tell the contractor when to expect payment. Unfortunately, only a few states have laws requiring that home-improvement contracts specify starting and completion dates.

The contract should allow for reasonable delay beyond the control of the contractor (such as a spell of bad weather that keeps roofers from doing their job), and most contractors will insist on a clause that gives them some leeway. But a contract that allows for delays "due to fire, strikes, war, riots, or other civil disturbances, governmental regulations or prohibitions, material or labor shortages, accidents, lockouts, force majeure, acts of God, adverse weather conditions, or any other cause or condition beyond the control of the contract" excuses almost every kind of foot-dragging. And it's inordinate work delays that are a major cause of consumer exasperation with contractors and other building professionals.

You can try to keep a contractor on schedule by including in the contract a sentence that says "All time limits in the contract are of the essence of the contract." You should note that this clause also obligates you to pay the contractor on schedule if the work is proceeding without delay.

Having that provision included in the contract, however, does not give you all the protection you need. You must also take action should the contractor fail to meet a deadline. If you ignore a missed deadline, you might be effectively waiving the "time is of the essence" provision and would then have no recourse if the contractor continued to delay completion of the work.

Hold-back clause. Another way to keep work delays to a minimum is to insert a clause allowing you to withhold final payment if the work slows down for no acceptable reason.

Resisting any pressure from the contractor, you should insist on making the final payment 30 days or so after the contractor finishes the work. That allows you time to live with the improvements—and gives the contractor an excellent incentive to come back to fix anything that wasn't completed or done properly the first time. A hold-back clause can also protect you against other problems, such as liens against your property. (Liens are discussed later in this section.)

A strong hold-back clause would say, "Final payment shall be due _____ business days after completion of work by contractor. Final payment may be withheld on account of defective work that is not remedied; claims filed; failure of the contractor to make payments properly to subcontractors or for labor, materials, or equipment; or unsatisfactory prosecution of the work by the contractor." (The last phrase, *unsatisfactory prosecution of the work,* is intended to cover unreasonable delays.)

Modifications. One contract the NCLC looked at said, "I have read the above contract. I agree to pay for any additional work not stated in the above contract." Avoid at all costs that kind of language. It gives the contractor the right to charge you for work you didn't want and may not need. The contract should make clear that you and the contractor must agree on changes in the job specifications—and the cost of the changes—before any extra work can be started. (Keep in mind that asking for changes once the work has begun can cost a lot more money than you might think.) A suitable clause would say: "This contract can be modified only if the homeowner and the contractor sign a later agreement that sets forth the changes agreed to. If there are any work modifications, the resulting costs or credits to the homeowner will be included in that agreement."

Schedule of payments. Most state laws that cover payment schedules specify only that the schedule shall be written into the contract. Some states' home-improvement laws may be more explicit. "Never let the contractor get ahead of you," as one official put it. For example, if you're having the attic insulated and finished, you might give the contractor a small down payment, with the balance to be paid in four installments, as follows: the first when materials are delivered, the second when the insulation is installed, the third when the attic needs only a coat of paint to be fully finished, and the final payment *after* the job is completed to your satisfaction.

Pacing the payment schedule to the work could be somewhat difficult if you've borrowed money to pay for your home improvements. If you've taken a loan to pay for the work, you should discuss with the bank how you can retain some control over the way the loan is disbursed. You could ask the bank, for example, to issue the loan in installments and to make the checks payable to you and the contractor. You might also be able to have the bank issue the payments according to the rate at which the contractor's work progresses. Or the bank could agree to give you the full amount of the loan as one lump sum, which you would then transfer to the contractor according to the payment schedule included in the contract for the work.

If you've arranged financing through the contractor, the payment schedule probably will require repayment in equal monthly installments. From a homeowner's standpoint, the most advantageous way to handle the payments for contractor financing is to specify a payment schedule that does not begin until after the work has been completed satisfactorily, but this type of provision may be difficult to arrange.

If the contractor is arranging financing for the improvements, the contract must conform to the federal truth-in-lending laws. Among other things, truth-in-lending gives you the right to cancel the contract without penalty for at least three business days after you sign it, if your house is being used as loan collateral. The law also requires that the contract spell out the interest rate in terms of the annual percentage rate (APR), the cash price plus the finance charge, and the amount of each payment.

Permits and variances. You may encounter contract language that says, "Buyer will identify boundary lines and be responsible for obtaining all necessary permits and zoning variations before commencement of work." It's better, we think, to give that task to the contractor. A suitable clause could say, "The contractor shall be responsible for all permits, fees, and licenses necessary for the execution of the work. Further, the contractor shall give all notices and comply with all laws, ordinances, rules, regulations, and orders of any public authority bearing on the performance of the work."

Protection against liens. Just because you've paid the contractor doesn't mean that the money has been passed on to subcontractors or suppliers. If a contractor fails to pay suppliers, they can slap a lien (called a *mechanic's* or *materialman's lien*) on your house. In effect, the lien gives suppliers the right to take your property (or some of it, at least) as payment. Some states require home-improvement contracts to include a warning about such liens. Mere notification is better than nothing, but you should try to get a contract that offers some protection as well. One way to do that is to hold up the final payment until the contractor offers evidence that no liens will be filed. Such a clause might say, "Final payment shall not be due until the contractor has delivered to the homeowner a complete release of all liens arising out of the contract, or receipts in full covering all labor, materials, and equipment for which a lien could be filed." A second option is to insist that the contractor post a bond that would protect you against any liens. Only a handful of states give homeowners the right to demand that kind of bond.

Liability coverage. A few states require contractors to carry insurance to cover liability, workers' compensation claims, or both. Whether required by law or not, it's advisable to deal with a contractor who carries adequate insurance. The contractor should offer proof of insurance and the contract should specify that you're protected if claims arise.

Warranty on the work. A contractor should be willing to back up the quality of the work in writing, by including a warranty in the contract. Some of the contracts the NCLC examined, however, included no warranties, while others offered only vague assurance: "Contractor will do all of said work in a good, workmanlike manner." A stronger clause would say, "The contractor shall remedy any defects that appear during the progress of the work. Contractor warrants the work to be performed under this contract to be free from defects in material and workmanship for a period of ____ years from the date of completion. This provision applies to work done by subcontractors as well as to work done by the contractor's employees."

Some contractors say that homeowners should expect at least a one-year warranty on the work. Still, the duration of the warranty will depend on the nature of the work and your ability to negotiate with the contractor. Manufacturers of shingles and siding, for example, generally give a warranty on their products for 15 years or more. A contractor should provide a commensurate warranty on the installation of those materials.

Be sure, too, that the contractor gives you the manufacturers' written warranties that accompany fixtures, appliances, or other products. Keep these documents on file.

Cleanup. If the contractor's form contract says, "Rubbish removal is homeowner's responsibility," have that clause changed. Insist on a clause that puts the burden of cleaning on the contractor. The standard language for such a clause requires the contractor to leave the premises in "broom-clean condition."

Credits and returns. Some of the contracts the NCLC examined included clauses that said, in essence, that some of the surplus materials remain the property of the contractor, with no mention of credit given to the homeowner. It's better, we think, to have a clause that reads, "The surplus material that can be returned to the supplier will be credited to the homeowner." If it's not going to be used, you shouldn't have to pay for it.

Cancellation rights. Some states and cities have laws that give you a cooling-off period (similar to the provisions of the federal truth-in-lending laws) that allows you a few days to review the contract and, if you wish, cancel it without penalty. If you're unsure of the law in your area, contact the local consumer protection agency or your state attorney general. If a cooling-off period is required, be sure the contract includes it. And don't allow the contractor to purchase materials or begin work until the cooling-off period ends.

Some contracts may also contain a "liquidated-damages" provision—in effect, a penalty clause if you decide to cancel after a cooling-off period expires. One such provision says, "Homeowner agrees that in event of cancellation of this contract before work is started, homeowner shall pay to contractor on demand 25 percent of the contract price as its stipulated damages for the breach." In other words, if you signed a contract for a $4,000 job and canceled four days later (after the cooling-off period had lapsed), it could cost you $1,000 to back out if the contractor succeeded in enforcing a liquidated-damages claim.

A provision that's much less damaging, from your point of view, would say, "The contractor is entitled to 5 percent of the cash price, but no more than $100, if the homeowner cancels this contract more than three business days after it was signed, but before materials have been delivered or work has started." But if the contractor tries to enforce a contract with a liquidated-damages provision, be sure to find out before you pay what your state laws specify can be collected under that provision. Some states have put a ceiling on the amount of damages a contractor can demand.

Keep in mind that a contractor can sue for actual damages if, at some point, you fail to meet your contractual obligations.

Some Further Precautions

Read all documents carefully; you may want an attorney to review the contract and any other forms before you sign them. In any event, do not sign a "certificate of completion" (or a document by any other name that says the work has been completed to your satisfaction) until the work has been finished and you have inspected it carefully. In some states, it is illegal for a contractor to have a certificate of completion signed in advance. In Florida, when the final payment becomes due, the general contractor is required to provide the homeowner with an affidavit indicating which subcontractors and suppliers have been paid in full.

Taking the time to find a reputable contractor and including a hold-back clause in the contract will help ensure that the contractor finishes the work on time and does the job properly. Still, there's always the risk that the contractor won't follow the specifications called for in the contract. The contractor may also cut corners, even though the workers appear to be doing their job correctly. Those shortcomings might not be revealed for weeks, or months—long after the contractor has been paid in full.

Granted, it's impractical to watch every move the work crew makes. But if the workers do something that doesn't look right, clear up the problem right away; don't wait until the job has been completed to raise questions or lodge complaints. You might also try one remodeling expert's method for keeping contractors on their toes: Take pictures of the work.

SELECTING THE CONTRACTOR AND NEGOTIATING THE CONTRACT (WORKSHEET)

You can refer to this form in deciding which contractor to select.* Use this form in conjunction with the preceding section on seeking professional help. Because you may want to get written bids on work to be done from a few contractors, parts of the form are designed for three sets of entries.

*In preparing the form, Consumers Union used as a starting point a checklist for choosing a contractor distributed by a public utility in Westchester County, New York.

I. Identify the job to be done.
Before you call a contractor for an appointment, define specifically the work you want done.

A. List your initial specifications for the job.

_____

_____

_____

_____

B. Include the kinds of material you want used.

_____

_____

_____

_____

C. Decide on your budget for the project. $_____

II. Ask for bids.
A. Get written bids on the job from at least three contractors. (You can locate contractors through friends, neighbors, people at local building supply centers, or your utility company.)

	Contractor 1	Contractor 2	Contractor 3
Name	_____	_____	_____
Address	_____	_____	_____
Phone	_____	_____	_____
Recommended by:	_____	_____	_____

B. Which modifications suggested by the contractors do you want to add to your original specifications?

Contractor 1	Contractor 2	Contractor 3
_____	_____	_____
_____	_____	_____
_____	_____	_____
_____	_____	_____

C. What is the rough cost estimate for the job?

	Contractor 1	Contractor 2	Contractor 3
	$_____	$_____	$_____

D. What are the contractors' financing terms and arrangements for payment?

	Contractor 1	Contractor 2	Contractor 3
	_____	_____	_____
	_____	_____	_____
	_____	_____	_____
	_____	_____	_____

E. How long will it take to do the work?

	Contractor 1	Contractor 2	Contractor 3
	_____	_____	_____
	_____ Start on	_____ Start on	_____ Start on
	_____ Completed by	_____ Completed by	_____ Completed by

III. Check the contractors' experience.

A. If you are not familiar with the contractors' professional background, you may want answers to the following questions.

	Contractor 1	Contractor 2	Contractor 3
Years as contractor?	_____	_____	_____
Years doing installations required for the job?	_____	_____	_____
Licensed (by state, county, or locality)?	_____	_____	_____

B. Names and addresses of previous customers for whom the contractors have done similar work.

Contractor 1	Contractor 2	Contractor 3
_____	_____	_____
_____	_____	_____
_____	_____	_____

C. Questions for previous customers. (If possible, visit homes of previous customers, look over the work and discuss the job with them.)

	Contractor 1	Contractor 2	Contractor 3
1. Was job started on time?	_____	_____	_____
2. Was job completed on time?	_____	_____	_____
3. Did actual cost exceed estimate? (If yes, why was there an overrun?)	_____	_____	_____
4. Was customer pleased with job?	_____	_____	_____
5. Was contractor willing to fix things done wrong the first time?	_____	_____	_____
6. Did the contractor leave premises in satisfactory condition?	_____	_____	_____
7. Did contractor supervise work?	_____	_____	_____
8. Other comments?	_____	_____	_____

D. Contact licensing bureau of area consumer protection office for answers to the following questions.

	Contractor 1	Contractor 2	Contractor 3
Number of complaints filed?	_____	_____	_____
Nature of complaints?	_____	_____	_____
How were complaints resolved?	_____	_____	_____

IV. Negotiating the contract.

Once you have decided on a contractor, include in writing all the details of the job you've discussed with the contractor.

A. Spell out the items to be covered in the job.

_____

_____

_____

_____

B. Specify brand names, colors, grades, styles, model numbers for appliances and materials.

_____

_____

_____

_____

C. Are the following clauses and provisions included in the contract?
   1. Starting date.
   2. Completion date.
   3. Hold-back clause (final payment to be made about 30 days after contractor finishes the work).
   4. Changes in job specifications (and cost of any changes) must be subject to mutual agreement and be made in writing.
   5. Schedule of payments tied to satisfactory work.
   6. Right of cancellation without penalty for at least three business days after signing contract.
   7. Specify interest rate in terms of annual percentage rate, cash price plus finance charge, and amount of each payment.

8. Contractor required to secure all necessary permits, fees, and licenses for the job and to do all work in compliance with applicable laws, ordinances, and regulations.
9. Contractor to deliver to homeowner a complete release of all liens arising out of contract prior to receipt of final payment (or to post a bond protecting homeowner against liens).
10. Contractor to provide homeowner with proof of insurance to cover liability, workers' compensation claims, or both (and to specify in contract that homeowner is protected if claims arise).
11. Contractor's warranty for work and material, which should also cover any work by subcontractors.
12. Contractor to leave premises in "broom-clean condition." Exterior of premises to be left in good condition.
13. Contract to specify credit or refund to homeowner for surplus materials.
14. Minimal penalty clause for cancellation following expiration of cooling-off period (try to avoid "liquidated-damages" provision).

V. Additional points.
1. Any architect's or engineer's drawings required for the job should be attached to the contract and cited in specifications.
2. The contractor should turn over to the homeowner all written manufacturers' warranties that accompany fixtures, appliances, or other products.
3. A homeowner should not sign a "certificate of completion" or similar document until work has been finished and the job has been carefully inspected (including inspection by a building inspector, if required).
4. A contractor may insist on using a preprinted form for the agreement. Such forms seldom include space for additions to the contract. A homeowner should thus be prepared to incorporate additional terms into the preprinted form by means of a separate memorandum. To ensure that the added page(s) become part of the preprinted form, the homeowner should write in on the form contract a statement along the following lines: "This agreement is subject to all the terms, representations, and promises stated on a separate memorandum dated [_____] and are hereby included as part of this agreement."

   The additional terms can then be written on a separate sheet of paper, prefaced by a heading similar to the one below.

Memorandum of terms of contract

Date _____

As stated in the contractor's printed agreement dated [_____], the sale of [*describe job to be contracted as stated in printed form*] by [*state contractor's name and address as in printed form*] is subject to the following terms and contractor's representations and promises that were specifically consented to as a basis for that agreement.

The homeowner should then simply itemize the terms and specifications that were part of the deal. If the memorandum includes more than one page, each page should be numbered in the following fashion: page _____ of _____ pages. The homeowner should make a copy of the memorandum for the contractor. The contractor should initial each page of the memorandum.

Appendix A

STATE ENERGY OFFICES

State energy offices can supply you with information concerning state laws and corresponding programs that are designed to conserve energy. The most common programs offered are:

- Incentive programs or other inducements to invest in energy-saving products
- Utility-rate reform, which includes special rate schedules for the poor and elderly, new rates designed to promote conservation, and rate reforms that apply only to a single utility
- Building codes and standards, which include conservation guidelines added to building codes, as well as standards for solar-energy equipment, insulation, and air conditioners and other appliances.

A list of the state and territorial offices follows.

Alabama
Chief
Department of Economics and
  Community Affairs
Energy Division
3465 Norman Bridge Road, Suite
  300
Montgomery, Alabama 36105-0939
(205) 284-8453

Alaska
Deputy Director, World
  Development
Division of Energy Programs
949 East 36th, Suite 403
Anchorage, Alaska 99508
(907) 563-1955

American Samoa
Director
Territorial Energy Office
Office of the Governor
Pago Pago, American Samoa 96799
(684) 699-1325

Arizona
Director
Arizona Energy Office of the
  Department of Commerce
1700 West Washington, 5th floor
Phoenix, Arizona 85007
(602) 255-3632

Arkansas
Director
Arkansas Energy Office
No. 1 State Capitol Mall, Rm.
    2C-105
Little Rock, Arkansas 72201
(501) 682-1370

California
Chairman
California Energy Commission
1516 Ninth Street
Sacramento, California 95814
(916) 324-3326

Colorado
Director
Colorado Office of Energy
    Conservation
112 East 14th Avenue
Denver, Colorado 80203
(303) 894-2144

Connecticut
Under Secretary for Energy
Office of Policy and Management
Energy Division
80 Washington Street
Hartford, Connecticut 06106
(203) 566-2800

Delaware
Energy Program Administrator
Division of Facilities Management
Energy Office
P.O. Box 1401, O'Neill Building
Dover, Delaware 19903
(800) 282-8616 (Delaware only)
(302) 736-5644

District of Columbia
Director
DC Energy Office
613 G Street, N.W., Room 500
Washington, DC 20004
(202) 727-1800

Florida
Director
Governor's Energy Office
214 South Bronough Street
Tallahassee, Florida 32399-0001
(904) 488-6764

Georgia
Director
Office of Energy Resources
270 Washington Street, S.W., Suite
    615
Atlanta, Georgia 30334
(404) 656-5176

Guam
Director
Guam Energy Office
P.O. Box 2950
Agana, Guam 96910
(671) 734-4452/4530/2723
    (overseas operator)

Hawaii
Planning and Economic
    Development Department
DBEB, Energy Division
335 Merchant Street, Room 110
Honolulu, Hawaii 96813
(808) 548-4150

Idaho
Bureau Chief
Bureau of Energy Resources
Department of Water Resources
1301 N. Orchard Street
Boise, Idaho 83720
(208) 334-7968

Illinois
Director
Department of Energy and Natural
   Resources
325 West Adams, Room 300
Springfield, Illinois 62704
(217) 785-2002

Indiana
Director
Division of Energy Policy
Indiana Department of Commerce
One North Capitol, Suite 700
Indianapolis, Indiana 46204-2288
(317) 232-8946

Iowa
Administrator
Division of Energy and Geological
   Resources
Iowa Department of Natural
   Resources
Wallace State Office Building
Des Moines, Iowa 50319
(515) 281-5145

Kansas
Energy Programs Supervisor
Energy Division
Kansas Corporation Commission
Docking State Office Building, 4th
   Floor

Topeka, Kansas 66612-1571
(913) 296-5460

Kentucky
Secretary
Kentucky Energy Cabinet
P.O. Box 11888
Lexington, Kentucky 40578-1916
(606) 252-5535

Louisiana
Executive Officer
Energy Division
Department of Natural Resources
P.O. Box 94396
Baton Rouge, Louisiana 70804-
   9396
(504) 342-4500/4593

Maine
Director
Office of Energy Resources
State House Station 53
Augusta, Maine 04333
(207) 289-3811

Maryland
Director
Maryland Energy Office
45 Calvert Street
Annapolis, Maryland 21401
(301) 974-3751

Massachusetts
Secretary
Executive Office of Energy
   Resources
100 Cambridge Street, Room 1500
Boston, Massachusetts 02202
(617) 727-4732

Michigan
Director
Office of Energy Programs
Public Service Commission
Michigan Department of
    Commerce
P.O. Box 30221
Lansing, Michigan 48909
(517) 334-6272

Minnesota
Director
Energy Division
Public Service Department
160 East Kellogg Boulevard
St. Paul, Minnesota 55101
(612) 297-4685

Mississippi
Director
Mississippi Department of Energy
    and Transportation
Dickson Building
510 George Street, Suite 300
Jackson, Mississippi 39202-3096
(601) 961-4733

Missouri
Director
Department of Natural Resources
Division of Energy
P.O. Box 176
Jefferson City, Missouri 65012
(314) 751-4000

Montana
Administrator
Energy Division
Department of Natural Resources
    and Conservation

1520 East 6th Avenue
Helena, Montana 59620-2301
(406) 444-6754

Nebraska
Director
Nebraska State Energy Office
P.O. Box 95085, 9th Floor
State Capitol Building
Lincoln, Nebraska 68509-5085
(402) 471-2414

Nevada
LEHEA Program
Welfare Division
2527 N. Carson Street
Capitol Complex
Carson City, Nevada 89710
(702) 687-4420

New Hampshire
Director
Governor's Energy Office
2½ Beacon Street, 2nd Floor
Concord, New Hampshire 03301
(603) 271-2711

New Jersey
Commissioner
Department of Commerce, Energy
    and Economic Development
Energy Division
20 State Street, CN 820
Trenton, New Jersey 08625
(609) 292-2444

New Mexico
Secretary
Department of Energy, Minerals
    and Natural Resources

408 Galisteo, Vellagra Building
Santa Fe, New Mexico 87503
(505) 827-7836

New York
Commissioner
New York State Energy Office
2 Rockefeller Plaza, 10th Floor
Albany, New York 12223
(800) 342-3722 (New York only)
(518) 473-4376

North Carolina
Director
North Carolina Department of
    Commerce
Energy Division
430 N. Salisbury Street
Raleigh, North Carolina 27611
(919) 733-2230

North Dakota
Director
Office of Intergovernmental
    Assistance
State Capitol, 14th Floor
Bismarck, North Dakota 58505
(701) 224-2094

Northern Mariana Islands
Energy Administrator
Office of Energy and Environment
P.O. Box 340
Saipan, Mariana Islands 96950
(670) 322-9229
Routing 1-60-PLUS 671

Ohio
Chief
Energy Conservation
Department of Development

30 East Broad Street, 24th Floor
Columbus, Ohio 43266-0413
(614) 466-6797

Oklahoma
Director
Division of Conservation Services
Corporation Comm.
302 Jim Thorpe Building
Oklahoma City, Oklahoma 73105
(405) 521-4467

Oregon
Director
Department of Energy
625 Marion Street, N.E.
Salem, Oregon 97310
(503) 378-4040

Pennsylvania
Executive Director
Pennsylvania Energy Office
116 Pine Street
Harrisburg, Pennsylvania 17105
(717) 783-9982

Puerto Rico
Director
Puerto Rico Office of Energy
Office of the Governor
P.O. Box 41089-Minillas Station
Santurce, Puerto Rico 00940-1089
(809) 721-4190

Rhode Island
Executive Director
Governor's Office of Energy
    Assistance
275 Westminster Mall
Providence, Rhode Island 02903
(401) 277-3370/6920

South Carolina
Director
Governor's Office
Division of Energy, Agriculture and
  Natural Resources
1205 Pendleton Street, 3rd Floor
Columbia, South Carolina 29201
(803) 734-0445

South Dakota
Director
Office of Energy Policy
217½ West Missouri Street
Pierre, South Dakota 57501-4516
(605) 773-3603

Tennessee
Director
Department of Economic and
  Community Development
Energy Division
320 6th Avenue North, 6th Floor
Nashville, Tennessee 37219-5308
(615) 741-6671

Texas
Director
Office of Energy
Office of the Governor
P.O. Box 12428
Austin, Texas 78711
(512) 463-1878

Utah
Director
Utah Energy Office
355 West North Temple
3 Triad Center, Suite 450
Salt Lake City, Utah 84180-1204
(800) 662-3633 (Utah only)
(800) 538-5428

Vermont
Commissioner
Conservation and Renewable
  Energy Unit
Public Service Department
State Office Building, 120 State St.
Montpelier, Vermont 05602
(802) 828-2321

Virgin Islands
Director
Virgin Islands Energy Office
Old Customs House
Frederiksted, St. Croix, U.S.V.I.
  00820
(809) 772-2616

Virginia
Director
Division of Energy
Department of Mines, Minerals and
  Energy
2201 West Broad Street
Richmond, Virginia 23220
(800) 552-3831 (Virginia only)
(804) 367-6851

Washington
Director
Washington State Energy Office
809 Legion Way S.E.
Olympia, Washington 98504-1211
(206) 586-5000

West Virginia
Fuel and Energy Office
Governor's Office of Community
  and Industrial Development
Building 6, Room 553
State Capitol Complex

Charleston, West Virginia 25305
(304) 348-4010

Wisconsin
Administrator
Division of State Energy and
    Intergovernmental Relations
Department of Administration
101 South Webster, 6th Floor
P.O. Box 7868
Madison, Wisconsin 53707-7868
(608) 266-8234

Wyoming
Energy Division Lead
Economic Development and
    Stabilization Board
Energy Division
Herschler Building, Third Floor
    East
Cheyenne, Wyoming 82002
(307) 777-7284

GOVERNOR-DESIGNATED STATE GRANTEES

The federal government has designated offices within the states to receive weatherization assistance funds from the Department of Energy (DOE). A list of these follows. (Some office locations are the same as those given for state energy offices in the list that precedes this one.)

For further information about the DOE weatherization program (WAP) for low-income people, contact the DOE directly. Address your inquiry to CE532, United States Department of Energy, Conservation and Renewable Energy, 1000 Independence Avenue, S.W., Washington, D.C. 20585.

Alabama
Division Chief
Community Services Division
Alabama Department of Economic
    and Community Affairs
3465 Norman Bridge Road,
    P.O. Box 250347
Montgomery, Alabama 36125-0347
(205) 284-8955

Alaska
Director
Division of Housing Assistance
Alaska Department of Community
    and Regional Affairs
949 E. 36th Avenue, Suite 400

Anchorage, Alaska 99508
(907) 563-1955

Arizona
Director
Arizona State Energy Office
1700 West Washington, 5th Floor
Phoenix, Arizona 85007
(602) 280-1300

Arkansas
Director
Office of Community Services
P.O. Box 1437/Slot 1330
Little Rock, Arkansas 72203
(501) 682-8650

California
Director
Office of Economic Opportunity
700 North 10th Street, Suite 272
Sacramento, California 95814
(916) 323-8694

Colorado
Executive Director
Department of Local Affairs
1313 Sherman Street, Room 415
Denver, Colorado 80203
(303) 866-2771

Connecticut
Commissioner
Department of Human Resources
1049 Asylum Avenue
Hartford, Connecticut 06105
(203) 566-7890

Delaware
Secretary
Department of Community Affairs
P.O. Box 1401
Dover, Delaware 19901
(302) 736-4456

District of Columbia
Director
D.C. Department of Housing and
   Community Development
1133 North Capitol Street, N.E.
Washington, D.C. 20002
(202) 535-1500

Florida
Secretary
Florida Department of Community
   Affairs

Housing and Community
   Development
2740 Centerview Drive
Tallahassee, Florida 32399-2100
(904) 487-3481

Georgia
Director
Office of Energy Resources
State of Georgia
270 Washington Street, S.W.
Atlanta, Georgia 30334
(404) 656-5176

Hawaii
Director
Office of Community Services
Department of Labor and Industrial
   Relations
335 Merchant Street, Room 101
Honolulu, Hawaii 96813
(808) 548-2130

Idaho
Supervisor
Idaho Department of Health and
   Welfare
State Economic Opportunity Office
450 West State Street
Boise, Idaho 83720
(208) 554-5730

Illinois
Director
Department of Commerce and
   Community Affairs
620 East Adams, 3rd Floor
Springfield, Illinois 62701
(217) 782-7500

Indiana
Acting Commissioner
Indiana Department of Human
    Services
251 N. Illinois Street, P.O. Box
    7083
Indianapolis, Indiana 46207
(317) 232-1147

Iowa
Administrator
Division of Community Action
    Agencies
Department of Human Rights
Lucas State Office Building
Des Moines, Iowa 50315
(515) 281-4420

Kansas
Administrator
Economic Opportunity Programs
Biddle Building, 1st Floor
300 SW Oakley Street
Topeka, Kansas 66606
(913) 296-2066

Kentucky
Acting Commissioner
Department for Employment
    Services
Cabinet for Human Resources
Commonwealth of Kentucky
275 East Main Street
Frankfort, Kentucky 40621
(502) 564-7015

Louisiana
Assistant Secretary
Department of Social Services
P.O. Box 44367, 1755 Florida
    Boulevard

Baton Rouge, Louisiana 70804
(504) 342-2297

Maine
Director
Division of Community Services
State House Station 73
Augusta, Maine 04333
(207) 289-3771

Maryland
Secretary
Maryland State Weatherization
    Program
45 Calvert Street
Annapolis, Maryland 21401
(301) 974-3176

Massachusetts
Secretary
Executive Office of Communities
    and Development
Leverett Saltonstall Building
100 Cambridge Street, Room 1404
Boston, Massachusetts 02202
(617) 727-7765

Michigan
Director
Michigan Department of Labor
Bureau of Community Services
P.O. Box 30015
Lansing, Michigan 48909
(517) 335-5953

Minnesota
Commissioner
Economic Opportunity Office
Department of Jobs and Training
Office of the Commissioner

390 North Robert Street
St. Paul, Minnesota 55101
(612) 296-3711

Mississippi
Director
Division of Energy and Community
  Services
Governor's Office of Human
  Development
State of Mississippi
301 West Pearl Street
Jackson, Mississippi 39203-3090
(601) 949-2038

Missouri
Director
Department of Natural Resources
Division of Energy
P.O. Box 176
Jefferson City, Missouri 65102
(314) 751-4422

Montana
Administrator
Economic Assistance Division
Department of Social and
  Rehabilitation Services
P.O. Box 4210, Capitol Station
Helena, Montana 59604-4210
(406) 444-4540

Navajo Nations
Director
Navajo Housing Services
  Department
P.O. Box 2396
Window Rock, Arizona 86515
(602) 871-6493

Nebraska
Director
Nebraska State Energy Office
9th Floor, State Capitol
P.O. Box 95085
Lincoln, Nebraska 68509
(402) 471-2867

Nevada
Administrator
Nevada State Welfare Division
Capitol Complex
2527 North Carson Street
Carson City, Nevada 89710
(702) 885-4128

New Hampshire
Director
Office of the Governor
Division of Human Resources
11 Depot Street
Concord, New Hampshire 03301
(603) 271-2611

New Jersey
Director
Division of Community Resources
New Jersey Department of
  Community Affairs
William Ashby Community Affairs
  Building
101 South Broad Street, CN800
Trenton, New Jersey 08625-0800
(609) 292-6212

New Mexico
Secretary
New Mexico Energy, Minerals, and
  Natural Resources Department
2040 South Paetteco

Santa Fe, New Mexico 87505
(505) 827-5860

New York
Director
Division of Economic Opportunity
New York State Department of
State
162 Washington Avenue
Albany, New York 12231
(518) 474-5700

North Carolina
Director
Energy Division
North Carolina Department of
Commerce
P.O. Box 25249
Raleigh, North Carolina 27611
(919) 733-2230

North Dakota
Director
Office of Intergovernmental
Assistance
State Capitol Building
Bismarck, North Dakota 58505
(701) 224-2094

Ohio
Deputy Director
Community Development Division
Ohio Department of Development
P.O. Box 1001
Columbus, Ohio 43266
(614) 466-2969

Oklahoma
Executive Director
Oklahoma Department of
Commerce

P.O. Box 26980
6601 Broadway Extension
Oklahoma City, Oklahoma 73116-
0980
(405) 521-3501

Oregon
Manager
Oregon State Community Services
207 Public Service Building
Salem, Oregon 97310
(503) 378-4729

Pennsylvania
Secretary
Department of Community Affairs
P.O. Box 155
Harrisburg, Pennsylvania 17120
(717) 787-7160

Rhode Island
Director
Governor's Office of Housing,
Energy and Intergovernmental
Relations
State House, Room 143
Providence, Rhode Island 02903
(401) 277-3370

South Carolina
Director
Office of the Governor
Division of Economic Opportunity
1205 Pendleton Street, 3rd Floor
Columbia, South Carolina 29201-
3713
(803) 758-3191

South Dakota
Secretary
Department of Social Services

Kneip Building
700 Governor's Drive
Pierre, South Dakota 57501
(605) 773-3766

Tennessee
Commissioner
Tennessee Department of Human
Services
Citizens Plaza Building
400 Deaderick Street
Nashville, Tennessee 37219
(615) 741-4964

Texas
Executive Director
Texas Department of Community
Affairs
P.O. Box 13166, Capitol Station
Austin, Texas 78711-3166
(512) 834-6010

Utah
Director
Utah Energy Office
355 West North Temple
3 Triad Center, Suite 450
Salt Lake City, Utah 84180-1204
(801) 538-5428

Vermont
Director
State Economic Opportunity Office
Agency of Human Services
103 South Main Street
Waterbury, Vermont 05676
(802) 241-2450

Virginia
Commissioner

Virginia Department of Social
Services
8700 Discovery Drive
Richmond, Virginia 23288
(804) 281-9046

Washington
Director
Department of Community
Development
9th and Columbia Building
Olympia, Washington 98504
(206) 753-4106

West Virginia
Director
Office of Economic Opportunity
1204 Kanawha Boulevard East
Charleston, West Virginia 25301
(304) 348-4010

Wisconsin
Secretary
Department of Health and Social
Services
Division of Community Services
1 West Wilson Street, P.O. Box
7851
Madison, Wisconsin 53707
(608) 266-3681

Wyoming
State Program Administrator
Department of Health and Social
Services
Division of Public Assistance and
Social Services
Hathaway Building, 3rd Floor
Cheyenne, Wyoming 82002
(307) 777-6068

Appendix B

**SELECTED RATINGS
OF HOME ENERGY USERS**

Listed in order of estimated quality; essentially similar models are bracketed and listed alphabetically. As published in a July 1990 report.

Specifications and Features

All: ● Are rated at 115 volts and 7.3 to 8.5 amps but can draw more under adverse conditions. ● Should be plugged into grounded outlet on 15-amp circuit protected by time-delay fuse or circuit breaker. ● Are designed for installation in double-hung window by two people. ● After proper installation, keep window from being raised. ● Have removable air filter. ● Use R-22 as refrigerant, a chemical with relatively low potential to harm environment. ● Have slinger ring to propel moisture extracted from room air onto condenser, to minimize dripping on very humid days.

Except as noted, all have: ● Adjustable vertical and horizontal louvers. ● Vent control to exhaust some room air. ● Three cooling speeds. ● Expandable side panels. ● Leveling provision. ● Power cord at least 70 in. long.

Key to Advantages

A—Pull-out filter can be reached for cleaning without removing front panel. Deep windowsills may block filter on **GE, Panasonic CW802JU,** and **Quasar,** which slides out from bottom.

B—Horizontal louvers can direct air flow downward.

C—Slide-out chassis makes installation safer and repair easier than with others.

D—External handle on top helps steady unit during installation.

E—Thermostat is calibrated in degrees Fahrenheit. (**Frigidaire** thermostat judged less accurate than others.)

F—Expandable side panels with metal framing; judged stronger than all-plastic panels.

G—Louvers can be closed to increase air thrust or vent room air to the outside.

H—Ventilated room air to the outside better than most.

I—Did not drip inside or outside during 4-hour high-humidity test.

J—Built-in on/off timer lets you set unit to start before you come home or run for a time then shut off automatically.

K—Has fresh-air vent and room-exhaust feature.

L—Has 5 fan speeds.

M—Power cord can be extended from left or right side of unit.

N—Installation instructions explain how to mount unit in window fitted with storm sash.

O—Signal light shines to show filter needs cleaning; other lights show operation in energy-saver mode.

P—L-shaped design minimizes height of unit in window.

Key to Disadvantages

a—Filter was hard to remove and clean because of screw holding grille in place.

b—Has fixed horizontal louvers.

c—Lacks leveling provision and does not have slide-out chassis; installation judged less safe than most. Optional mounting bracket available, $70.

d—Lacks guard over outdoor cooling fins.

e—Guard on outdoor cooling fins judged inadequate.

f—Harder to install than most; side curtains require assembly.

g—Harder to install than most; side panels must be cut to fit window.

h—Dripped slightly from interior side after 4-hour high-humidity test.

i—Selector-switch positions judged harder than usual to decipher.

j—Lacks vent to send room air outside, a minor drawback.

k—Unit should not be picked up by side-panel frames; bottom of side-panel support can easily slip out of frame, dropping unit to floor.

l—Power cord only about 50 inches long.

m—Plastic grille inside room has sharp corners.

n—Unreinforced foam-plastic duct was very vulnerable to breakage when moving unit in and out of chassis (manufacturer does warn of problem).

o—To leave energy-saver mode, unit must be turned off or go first to fan mode, a minor inconvenience.

p—Change in noise levels was greater than most when compressor cycled on or off at low fan speed.

q—Thermostat has no numerical markings.

Key to Comments

A—Instructions include information on through-the-wall installation.

B—Came with condensate drain kit (kit for **Friedrich** available as option, $19).

C—Manufacturer advises oiling fan motor periodically.

D—Has two cooling speeds; judged adequate.

E—Has no fiberglass insulation in output duct.

F—Has dual side discharge.

G—Unit's top frame/handle is 2 inches high, may prevent closing window with external handles.

H—Required partial disassembly before installation to remove shipping block.

Better ◄——————► Worse

| Brand and model | Price | Cooling capacity, Btu/hour | Energy efficiency rating | Moisture removal | Thermostat performance | | | | | Noise | | | | Advantages | Disadvantages | Comments |
					Regular	Energy-saver	Brownout conditions	Temperature uniformity	Directional control	Indoors, high	Indoors, low	Outdoors, high	Outdoors, low			
Panasonic CW801HU	$444	8000	9.1	2.7 pt./hr.	●	—	●	◐	◐	●	○	○		A,D,F,P	e,f,j	D,E
General Electric AME08FA	447	8000	9.5	2.5	●	●	●	◐	◐	○	◐	◐	◐	A,B,C,F,J	e,f,n	B,E,H
Panasonic CW802JU	457	8000	9.5	2.5	●	●	●	◐	◐	○	◐	◐	◐	A,B,C,E,F,J,N	e,f,n	B,E,H
Quasar HQ2082DW	404	8000	9.5	2.5	●	●	●	◐	◐	○	◐	◐	◐	A,B,C,E,F,J,N	e,f,n	B,E,N
Teknika AK84E	412	8000	9.7	2.3	●	—	●	◐	●	○	○	○	○	A,B,C,F,M	e,f	E,H

Brand and model	Price	Cooling capacity, Btu/hour	Energy efficiency rating	Moisture removal	Regular	Energy-saver	Brownout conditions	Temperature uniformity	Directional control	Indoors, high	Indoors, low	Outdoors, high	Outdoors, low	Advantages	Disadvantages	Comments
Frigidaire A08LH8N	446	7500	8.7	2.4	◉	◉	◉	○	◐	●	○	●	◡	A,D,E,G,H,J,N,O	d,h	E,G
White-Westinghouse AC088N7B	364	7500	8.7	2.4	◉	◉	◐	○	◐	●	○	●	◡	A,D,E,G,H,N,O	d,h	D,E,G
Emerson Quiet Kool 8DC73	417	8200	9.0	2.4	○	◉	◐	◐	◐	○	◐	◡	◡	A,D,E,I,N	—	D,E
Friedrich SS08H10A	566	8200	10.5	1.4	◐	○	◐	◐	◐	○	○	◡	◡	B,C,I,K,L,M	g,o	A,C
Carrier AGA1081	430	8000	9.0	2.7	◐	—	◐	◐	◐	◐	◐	◡	◡	A,B,C,N	f,p,q	A,E
Whirlpool ACQ082XW	422	8000	9.0	1.7	○	—	◐	◐	◐	◐	◐	◡	◡	A,B,C,F,H	d,h,l	E,F
Amana 9P2MB	390	8600	9.0	3.0	◐	—	◐	◐	◐	◐	◐	◡	◡	—	b,d,m	C
Hotpoint KVD08FA	390	7900	8.7	2.1	○	○	○	◉	○	○	◐	●	◡	—	—	—
Sears Kenmore 77088	417	8000	9.7	2.4	○	◐	◉	◐	◐	◐	◐	◡	◡	F,G,H	a,c,d,k	C
Sharp AF-808M6	439	8100	9.0	2.2	◉	○	○	◐	◐	○	◐	◡	◡	A,C	d,f,l	B,E
Airtemp C2R08F2A	377	8000	9.0	2.8	◐	—	⓵	◐	◐	◐	◐	●	●	—	b,p	
Climatrol M2R08F2A	444	8000	9.0	2.8	◐	—	⓵	◐	◐	◐	◐	●	●	—	b,i,p,q	
Fedders A2R08F2A	410	8000	9.0	2.8	◐	—	⓵	◐	◐	◐	◐	●	●	—	b,p	

⓵ A very wide range of performance was noted among different samples of these essentially similar models.

Ratings of 6000–7000 Btu Room Air Conditioners

Listed in order of estimated quality. Except where separated by bold rules, differences between closely ranked models were slight. As published in a July 1989 report.

Specifications and Features

All: ● Can be run on 15-amp, 120-volt circuit protected by circuit breaker or time-delay fuse. ● Should be used only with grounded outlet. ● Are rated at 5.6 to 7.5 amps but may draw more current under adverse conditions. ● Should be installed by more than one person.

Except as noted, all: ● Are designed for installation in double-hung window. ● Have adjustable vertical and horizontal louvers. ● Have three cooling speeds. ● Dripped heavily during high-humidity test. ● Have convenient controls. ● Are mounted with windowsill support bracket or slide-out chassis. ● Have leveling adjustment. ● Include expandable side panels with plastic framing. ● Have guard to protect outdoor cooling fins.

Key to Advantages

A—Has framed pull-out filter, accessible without removing front panel. Deep windowsill may obstruct removal on **GE** and **Panasonic.**
B—Has framed filter.
C—Horizontal louvers can be adjusted downward.
D—Slide-out chassis makes installation safer than most.
E—Has handle to steady unit during installation.
F—Thermostat calibrated in degrees; judged fairly accurate.
G—Expandable side panels have metal framing; judged stronger than plastic.
H—Louver(s) can be closed to increased air thrust or ventilation.
I—Side panels lock in place without damaging window sash.
J—Ventilated better than most.
K—Did not drip during high-humidity test.
L—Built-in on/off timer.
M—Slide-out chassis easy to use.

Key to Disadvantages

a—Filter was harder than most to remove and clean.
b—Fixed horizontal louvers.
c—Lacks sill support bracket or slide-out chassis; leveling provision provides sill support.
d—Lacks leveling provision.
e—Lacks guard over outdoor cooling fins.
f—Guard on outdoor cooling fins judged inadequate.
g—Requires $47 kit for window installation.
h—Side curtains require assembly.
i—Installation judged less safe than most.
j—Harder to install than most.
k—Controls judged more difficult to operate than most.
l—Vent feature more difficult to operate than most.

Key to Comments

A—Designed for through-the-wall installation.
B—Has provision for condensate drain line.
C—Manufacturer recommends periodic oiling of fan motor.
D—Has two cooling speeds; judged adequate.
E—Monitor indicates when filter needs cleaning.
F—Nonfiberglass insulation in output duct.
G—Motor oscillates vertical louvers.
H—Dripped slightly during high-humidity test.

(continued)

Ratings of 6000–7000 BTU Room Air Conditioners Continued

● Better ◖——○——◗ ● Worse

Brand and model	Average price (range)	Cooling capacity, Btu/hour	Energy efficiency rating	Thermostat performance					Noise				Advantages	Disadvantages	Comments
				Regular	Energy-saver	Brownout conditions	Temperature uniformity	Directional control	Indoors, high	Indoors, low	Outdoors, high	Outdoors, low			
Emerson Quiet Kool 6CC53	$367 ($300–440)	6300	9.5	●	◐	◐	○	◐	○	◐	◐	○	A,E,F	b	F,H
General Electric AME06LA	368 (279–469)	6000	9.5	◐	◐	◐	○	◐	◐	○	◐	○	A,C,D,G,I,L	d,f,h,i	B,F,H
Panasonic CW-601JU	390 (350–430)	6000	9.5	◐	◐	◐	○	◐	◐	○	◐	○	A,C,D,G,I,L	d,f,h,i	B,F,H
Sears 74065	417 (339–449)	6000	9.4	◐	◐	◐	◐	◐	◐	◐	◐	○	B,D,H,J,K,M	g,j	A,B,C
Carrier 51ZMA7061	375 (300–479)	6100	9.2	○	◐	●	◐	●	◐	○	◐	○	A,C,D,G,L	d,h,k	C,F
Hotpoint KQM06LB	383 (320–450)	6450	9.1	●	—	○	◐	◐	◐	◐	◐	○	—	—	D
Amana 7P2MA	327 (279–400)	6650	9.5	◐	—	◐	◐	◐	○	◐	◐	○	—	b,e	C
White-Westinghouse AC079M7B	405 (350–460)	7000	9.3	●	◐	◐	◐	○	◐	○	●	○	A,E,H	c,e	D,E
Whirlpool ACP602XT	353 (299–360)	6000	9.0	◐	—	◐	◐	◐	○	○	◐	○	G	a,b,d,e,i	C,H
Sharp AF-608M6	361 (279–430)	6500	9.5	◐	◐	◐	◐	○	○	○	○	○	—	d,f,i,l	F,H
Friedrich SP06G10D	478 (385–529)	6500	9.4	○	○	○	○	◐	○	○	◐	○	C,D,G,I	d	C,J
Gibson AL07A6EV	376 (300–453)	7000	9.3	●	◐	◐	◐	●	◐	○	◐	○	A,H,I,J	e	G
Airtemp CLR07F2K	305 (269–339)	7000	9.0	◐	—	●	◐	○	◐	○	◐	○	—	b	
Fedders ALR07F2J	390 (285–459)	7000	9.0	◐	—	●	●	○	◐	○	◐	○	—	b	
Climatrol MLR07F2J	327 (325–330)	7000	9.0	◐	—	●	●	○	◐	○	◐	○	—	b	D

Ratings of Clothes Dryers

Listed by types; within types, listed in order of estimated quality. As published in an October 1989 report.

Specifications and Features

Except as noted, all have: ● At least two heat settings. ● End-of-cycle signal. ● Baked-enamel finish. ● Door that opens to right. ● Venting from rear, either side, or bottom. ● Well-calibrated automatic dryness control. ● At least 60 minutes of timed dry.

Key to Advantages

A—Added tumbling time on permanent-press cycle. Separate "refresher" cycle with **Ward's.**
B—Drum light.
C—Loudness of end-of-cycle signal can be adjusted.
D—End-of-cycle signal can be turned off.
E—Signals when lint filter is full.
F—Console has fluorescent display.
G—Lighted display.
H—Comes with rack for drying items without tumbling.
I—Comes with exterior rack for hanging clothes.
J—Door mounted high enough to clear tall basket.
K—Oversize opening makes loading and unloading easier.
L—Porcelain-coated top.
M—Nonglare control panel.
N—Plastic leveling legs make dryer relatively easy to move.
O—Porcelain-coated drum.
P—Fabric-softener dispenser.
Q—Lint filter judged more effective than others.

Key to Disadvantages

a—No drum light.
b—No end-of-cycle signal.
c—Start switch is part of cycle selector, which only turns one way.
d—No timed dry cycle with heat.
e—Relatively small dryer opening; large items have to be loaded one at a time.
f—Has buttons for three heat settings but gives only two.
g—Cycle selector only turns one way.

Key to Comments

A—Timed dry adjustable in 10-minute increments.
B—During added tumble, drum turns for 10 to 30 seconds every 5 minutes.
C—Lint filter pulls up from top of cabinet.
D—Door opens downward.
E—Door opens to left.
F—Vents from rear only.
G—Vents from rear, left, or bottom.
H—Vents from rear, right, or bottom.
I—Delicate time cycle has about 20-minute cooldown; also labeled "air fluff."
J—Continuous heat range control on all cycles.
K—Single heat range in auto mode.
L—Has less than 60 minutes of timed dry.
M—Optional drying rack available from manufacturer or distributor.
N—End-of-cycle buzzer sounds for 5 minutes before cycle ends.
O—Must remove lint filter to use drying rack—can be hazardous if forgotten.
P—Auto and "hi" heat are same setting.
Q—Delicate performance shown reflects auto dry. Manufacturer suggests timed dry.
R—Permanent-press performance shown reflects regular auto dry. Manufacturer suggests timed dry.
S—Single-control dryer. Fabric-cycle selector automatically sets dryer temperature.
T—Although this model has been discontinued and is no longer available, the information has been retained to permit comparisons.

GUIDE TO THE RATINGS

1 Price. An estimated average.

2 Dimensions. Rounded to next highest quarter inch. Height of cabinet includes control panel. Depth excludes handles. Allow more space to accommodate vent piping, depending on installation requirements.

3 Depth, door open. Total depth with door open 90 degrees, an important number if space is tight.

4 Drum volume. The largest contain about 6½ cubic feet; the smallest, about 5 cubic feet. The larger the drum, the more easily it can handle large, bulky laundry such as comforters or parkas.

5 Mixed-load drying. How well the automatic-drying setting handled a **small** (6-pound) load of underwear, hand towels, shirts and shorts, or a **large** mixed load—12 pounds of towels, jeans, and shirts. The better models dried the clothes cool to the touch, allowed us to either dry a load thoroughly or leave it damp for ironing, and turned off promptly when the clothes reached the degree of dryness selected.

6 Permanent-press. Based on 6-pound loads of permanent-press shirts, with machines set on automatic, permanent-press cycles. The judgment includes uniformity of drying (at "more dry"), how long it took the dryer to realize the load was dry, whether the dryer could leave clothes slightly damp, and how close to room temperature the load finished up.

7 Delicate. How well the automatic-drying setting handled a 3-pound load of nylon panties and pajamas, plus nightgowns and blouses. Most machines ran too hot for those items—up to 140°F for the gentlest machine, and 180° for the harshest. Using a short, timed setting would be kinder to delicate items, as some manufacturers recommend.

8 Convenience. Positive attributes are dials that turn in both directions, a versatile range of temperature and cycle settings, end-of-cycle buzzers, and drum lights. Other design pluses cover ease of loading and comfortable handles. The moisture-sensing models were judged independently of the temperature-sensing models, so as not to unfairly penalize the more rudimentary machines.

● ◖ ○ ◗ ●
Better ◄——————► Worse

Brand and model	Price	Dimensions, HxWxD, inches	Depth, door open, inches	Drum volume	Small	Large	Permanent-press	Delicate	Convenience	Advantages	Disadvantages	Comments
Electric models (moisture-sensing)												
KitchenAid KEYE 800	455	42½x29x28¾	42½	●	●	●	◖	○	◖	A,B,E,F,G,K,L,N,P,Q	—	A,B,C,D,F,M,T
Sears 69921	437	43x27x28	42¼	●	◖	●	○	○	◖	A,B,C,D,E,H,I,J,K,N,Q	—	B,D,J,O,T
Electric models (temperature-sensing)												
Whirlpool LE5800XS	344	42½x29x26¼	43½	◖	○	◖	○	○	◗	A,E,N,Q	—	B,C,D,F,J,M
Maytag DE312	393	43¾x28½x27	47	◖	○	○	◖	◗	◖	L,Q	d	E,G,S,T
Frigidaire DEDF	292	44x27x26½	48¾	○	◖	○	○	○	○	—	—	I,K,Q,S
Montgomery Ward 7308	282	43½x27x25½	47¼	○	○	○	◖	○	○	A	—	—
Kelvinator DEA500F	298	43½x27x26½	48¾	○	◖	○	○	○	◗	—	b	I,K,Q,S,U
Sears 68621	299	43x29x28	42	◖	○	○	○	○	◗	N,Q	e	C,D,F,L,M,S,T
Magic Chef YE20-3	328	43¾x27x28	50¼	●	○	○	○	○	◗	A,B,J,K,Q	c	M
Admiral DE20F-5	396	43¾x27x28	50¼	●	○	○	○	○	◗	A,J,K,Q	a,c	M,P,T
Amana LE2400	345	42x27x28	47¼	◖	○	○	○	○	◗	A	a,c,f	M,N,U
White-Westinghouse DE500K	280	43¾x27x27	49¼	○	○	○	○	○	◗	M	b	I,K,Q,S
Speed Queen NE4513	335	43x27x28	47¼	◖	○	○	○	○	◗	—	a,b	M,R,U
Hotpoint DLB2450B	268	42½x27x25	51	○	◖	◗	◖	○	◗	J,O	a,b	M
General Electric DDE6350G	313	42½x27x25	51	○	◗	◗	◖	○	◗	J,O	a,b	M
Gas models (temperature-sensing)												
Maytag DG312	453	43¾x28½x27	47	◖	◖	◖	○	◖	◖	L,Q	d	E,G,S,T
Whirlpool LG5801XS	381	42½x29x26¼	43½	◖	◖	◖	○	○	◗	A,E,N,Q	—	B,C,D,F,J,M
White-Westinghouse DG500K	315	43¾x27x27	49¼	○	○	○	◖	○	◗	M	b	H,I,K,Q,S
Sears 78621	331	43x29x28	42	◖	○	◖	◖	◖	◗	N,Q	e	C,D,F,L,M,S,T
Speed Queen NG4519	402	43x27x28	47¼	◖	◖	◖	○	○	◗	—	a,b,g	H,M,R,U
General Electric DDG6380G	351	42½x27x25½	51½	○	◖	◖	◖	◖	◗	J,O	a,b	G,M

Ratings of Dishwashers

Listed in order of estimated quality, based on washing ability, energy efficiency, convenience, safety, and other factors. Except where separated by bold rules, closely ranked models differed little in quality. As published in a May 1990 report.

❶ **Price.** Estimated average retail prices.

❷ **Washing ability.** Results with normal/regular cycle (or closest equivalent two-wash cycle) on test loads smeared with various foods. Each machine washed six loads with 140°F water, six loads with 120° water. Each load was inspected afterward to assess its cleanliness. Judgments here summarize results for both water temperatures. The machines generally didn't get loads quite as clean with 120°F water, even if the dishwasher's heater compensated by boosting the water temperature for some parts of the cycle.

❸ **Energy efficiency.** These scores take into account the direct use of electricity for washing and heated drying, along with the cost of heating the water in the water heater and within the dishwasher.

❹ **Noise.** Each machine was run in a plywood enclosure. A given machine might sound better or worse in your kitchen, but the relative standings should remain the same.

❺ **Cycle time.** Using cooler water often lengthens washing time because most machines need extra time to heat the water.

Specifications and Features

All: ● Require about 34½x24x24 inches (HxWxD) for installation and 24 to 26 inches clearance in front to open door fully. ● Have at least heavy, normal, and light wash cycles and rinse-and-hold cycle or equivalent. ● Let you select no-heat drying with any wash cycle. ● Have rinse-conditioner dispenser. ● Have heating element under lower rack. ● Have reversible front panel(s) for choice of colors.
Except as noted, all: ● Filters, when provided, are essentially self-cleaning. ● Have porcelain-coated tub interior. ● Have dial or display that indicates progress through a cycle. ● Have one full- or two half-size flatware baskets. ● Require 120-volt, 15-amp circuit.

Key to Advantages

A—Has systems monitor that displays a few malfunctions.
B—Has systems monitor with numerous displays for malfunctions or status of cycle.
C—Has hidden touchpad to lock controls from children.
D—Timer dial or digital display shows minutes left in cycle.
E—May be set for delayed start up to 6 hours (**KitchenAid** and **GE,** up to 9 hours; **Sears,** up to 12 hours).
F—Upper rack can be adjusted up and down or tilted to change overhead clearance.
G—Bottom of upper rack is terraced with steps to accommodate oversize plates in lower rack.
H—Accepts larger dinner plates (12 inches or more) better than most others.
I—Has no water tower to intrude into main rack; judged to provide more flexibility for loading pots, bowls, and larger items.
J—Accepts very tall glasses better than most.
K—Upper rack (lower rack on **Jenn Air** and **Maytag**) has one or two fold-down sections for two-tier stacking of cups and squat glasses.
L—Upper rack has pickets for two-tier stacking of cups and squat glasses.
M—Upper rack has adjustable or folding divider props.
N—Has extra covered basket for small items.
O—Has extra utensil basket.
P—Upper rack has removable dish stand and two loose lattices to hold glasses or cups.
Q—Flatware basket(s) has covered sections for small items.
R—Flatware basket has carrying handle.
S—Automatically drains water for extra protection against accidental overfilling and flooding.
T—Stainless-steel interior and door liner.
U—Solid-plastic interior and door liner.
V—Solid-plastic door liner.
W—Chassis has wheels or "shoes" to facilitate moving during installation or servicing.
X—Smooth control-panel surface easy to clean.

Key to Disadvantages

a—Did not dry glassware as well as most others (**KitchenAid,** with 120°F water; **Sears,** with 140°F water).
b—Did not dry flatware as well as most others (**Caloric,** with 120° water; others, at both water temperatures).
c—Timer isn't accessible: Canceling a cycle takes a minute or two. **GE** also dumps its detergent when cycle is canceled.
d—On the tested sample, it was difficult to avoid shortening first fill when advancing the timer dial to start.
e—Has signal lights rather than timer; gives only rough indication of progress through the cycles.
f—Timer dial does not give a clear indication of when cycle is finished.
g—Cannot take glasses as tall as most others.
h—Filter required regular cleaning during CU's tests.
i—Certain areas in tub and door occasionally required cleaning during CU's tests.
j—Flatware basket lacks carrying handle.
k—On the tested samples, lower rack moved somewhat stiffly, dragging along tub sides.
l—Door latch moved stiffly on the sample.
m—External fiberglass insulation blanket lacked protective sheath; may fray if not handled carefully during installation.

Key to Comments

A—Has electronic timer; even so, you can cancel cycle quickly.
B—Regular cycle includes only one wash phase, not two as in most machines, or was otherwise limited. CU ran tests using heavy wash to get a more comparable wash-and-rinse cycle.
C—Rack design is opposite the norm: Upper rack holds larger plates and pots, lower holds cups and glasses.
D—Has removable flatware rack mounted inside door.
E—Some water may splash out if you open door quickly during wash cycle.
F—Requires 20-amp circuit.
G—Manufacturer recommends minimum water temperature of 140°F, but overall washing performance suffered only slightly with 120° water.
H—Electronic controls.
I—Push-button controls.
J—Lacks super/pots-and-pans cycle.
K—Delayed start.
L—Water-heating switch.
M—Super/heavy cycle.
N—Delicate/china/crystal cycle.
O—Convertible or portable version for the **Caloric** is DCS418; for the **Frigidaire,** it's DW4400D; for the **Maytag,** it's WC702; for the **Sears,** it's 17485.
P—According to the manufacturer, model has been replaced by an essentially similar model.
Q—Although this model has been discontinued and no longer is available, the information has been retained to permit comparisons.

Better ←——————→ Worse

Brand and model	① Price	② Washing ability China	Flatware	Glasses	③ Energy efficiency, 140°F/120° water	④ Noise	⑤ Cycle time, 140°F/120° water	Water used	Advantages	Disadvantages	Comments
KitchenAid KUDS22ST	$718	●	●	◐	◐/●	◐	75/80 min.	9½ gal.	A,D,E,F,I,M,Q,X	—	A,H,K,L
Magic Chef DU110CA	451	●	●	◐	○/●	◐	80/110	11½	E,G,R,U	f	L,M
Maytag WU702	549	●	●	◐	◐/●	○	75/80	11½	I,J,K,M,Q,W	k,m	C,E,F,L,O,P
Jenn Air DU476	572	●	●	◐	◐/○	○	80/100	12	I,J,K,M,Q,W	k,m	C,E,F,J,L,M,P
KitchenAid KUDC220T	439	●	●	◐	○/●	○	80/80	12	I	a,m	L
General Electric GSD2800L	552	◐	●	○	◐/●	◐	80/80	10	B,C,D,E,G,J,K,Q,R,S,U,X		A,H,K,N
Sears 16985	619	●	●	◐	◐/●	◐	70/85	10½	B,C,D,E,F,J,K,M,N,O,Q,S,V,W,X		A,H,K,L,Q
Sears 16695	374	●	●	◐	◐/○	◐	70/115	10½	K,N,O,Q,V,W	a,d	L,P
General Electric GSD1200L	475	◐	●	○	○/●	◐	85/100	10½	G,J,K,Q,R,U	c	I,N,P
Frigidaire DW4500F	355	◐	●	○	◐/●	◐	70/70	11½	E,F,G,H,K,L,Q,R,V	i	B,K,L,M,O
Whirlpool DU8900XT	449	◐	○	◐	◐/○	◐	75/110	10½	E,F,O,V,W	b	B,D,K,L,M
Whirlpool DU9400XT	530	◐	○	◐	◐/●	◐	70/105	10½	C,E,F,V,W,X	b,e	A,B,D,H,K,L,M
Thermador TD5500	755	◐	◐	◐	○/●	○	80/80	12	P,Q,T	c,e,g	I
White-Westinghouse SU550J	314	◐	◐	○	○/●	◐	75/75	11½	D,G,L,R,V	i,m	B,M,Q
Caloric DUS406	300	◐	◐	○	○/○	○	75/85	11½	N,V	b,g,j	B,H,L,M,O
Hotpoint HDA2000G	380	○	◐	◐	◐/●	◐	75/85	10½	B,D,H,Q,R,U,X	—	A,H,N,P
Sears 16485	299	◐	●	◐	○/●	◐	75/75	11½	I,K,N,O,V	g,l	G,O,P
General Electric GSD600L	305	○	◐	◐	◐/●	◐	75/75	11½	H,R,U	—	P
Hotpoint HDA950G	340	○	◐	◐	◐/●	◐	70/80	10	H,R	b,g,m	L,P
Whirlpool DU7400XS	314	○	○	◐	◐/●	◐	60/60	10½	V,W	b,h	D,G

Ratings of Electric Blankets and Mattress Pads

Listed by types. Within types, listed in order of estimated quality. Models judged equal in overall quality are bracketed and listed in order of increasing price. As published in a November 1989 report.

Specifications and Features

All: ● Are 120-volt AC and UL-listed. ● Have illuminated controls with on/off switch (one control for twin, two for queen). ● Are machine-washable. Weigh less than 4 pounds (twin) or 6 pounds (queen).

◉ ◓ ○ ◒ ● Better ←——→ Worse

Brand & model	Price, twin/queen	Overall score (0–100)	Heating performance	Control visibility (top/side)	Control noise	Wear
Electric blankets						
Northern Electric Slumber Rest Genesis 5100/5104	$92/150		●	○/◐	○	◐
Northern Electric Sunbeam Concept 4670/4674	60/90		●	●/◐	●	◐
Northern Electric Sunbeam Warmcrest 3704 [1]	—/65		●	●/○	○	◐

(continued)

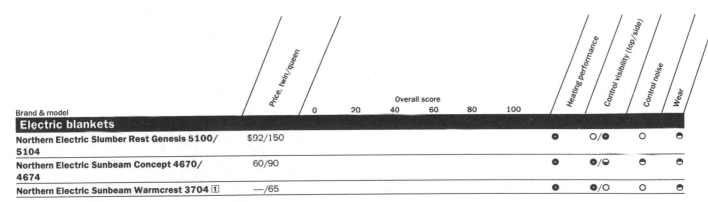

Brand & model	Price, twin/queen	Overall score	Heating performance	Control visibility (top/side)	Control noise	Wear
Electric blankets						
Sears Matchmate III Cat. No. 7100/7103	35/65		◑	◑/○	○	◓
Northern Electric Sunbeam Omni 5280-003/5284-003	65/105		◑	◒/◑	○	◓
Fieldcrest Cannon Ultimatic 18413/18483	65/115		○	◒/◑	◑	◒
Sears Smart Cat. No. 7150/7153	70/100		◓	◒/●	◑	◓
Northern Electric Sunbeam Warmcrest 3700 ②	—/45		◓	◑/○	○	◓
Fieldcrest Cannon St. Mary's Fairlane 07317/07387	25/50		○	◑/◒	●	○
Fieldcrest Cannon St. Mary's Celebrity 08117/08187	30/60		○	◑/◒	●	○
Fieldcrest Cannon Signature Compliments 04513/04583	45/65		○	◑/◒	●	○
Fieldcrest Cannon Fieldcrest 17113/17183	40/60		○	◒/◑	◒	◒
Electric mattress pads						
Sears Cat. No. 8265/8268	50/80		○	●/◑	○	◓
Northern Electric Sunbeam Fleece Bedwarmer 2060/2063	70/110		○	◑/○	◓	◒
Perfect Fit Automatic Bedsack 76806-2930/76806-912933	50/75		●	◒/◑	○	◒
Fieldcrest Cannon St. Mary's The Warm Up 06814/06884	25/60		◒	◒/●	◑	◓
Northern Electric Sunbeam Bedwarmer 2410/2413	40/70		◒	◑/○	○	◒
Sears Cat. No. 8245/8248	34/54		◒	●/○	○	◒
Chatham Restwarmer Lifetime 733401/2152	48/76		●	◑/◒	◑	○

① Queen size. ② Twin size.

Listed by size and by type of heat transmission, convection or radiation. Within types, listed in order of estimated quality. As published in a November 1989 report.

GUIDE TO THE RATINGS

❶ Price. Manufacturer's suggested retail. + means shipping is extra.

❷ Type/Dimensions. Any electric heater transforms all the electrical energy it draws into heat. But designers have found many ways to work the transformation—sometimes in a compact package, sometimes in a unit that's tall or long. **Coil** and **Wire** heaters use a fan to blow air over a heated metal coil or wire. Ceramic heaters **(Cer)** use a fan to blow air around or through a special conductive ceramic material. **Water** indicates water-filled baseboard heaters. **Calrod** indicates a baseboard model heated by an element similar to those on most electric ranges. **Oil/B** is

a baseboard model using a sealed, oil-filled tube. **Steam** is an upright heater using a sealed steam-recirculating system and a blower. **Oil/Up** is a sealed upright radiator filled with electrically heated oil. **Oil/wire** is an oil-filled upright radiator supplemented with a wire element and fan. Ribbon heaters **(Rib)** use a metal ribbon to radiate heat, supplemented with a fan. Most are rectangular boxes but two are baseboard heaters **(B)**. **Quartz** heaters can be either fan-driven or not and use incandescent coils packed inside quartz tubes. The tubes are oriented upright **(Up)** or horizontally **(H)**.

❸ Safety. While many of these heaters have a tip-over switch that will shut them off if they fall over, some can still pose a stability problem. In determining overall safety, fire safety is a primary concern. The "not acceptable" units ignited a piece of terrycloth draped over the grille in a matter of minutes.

❹ Performance. Temperature regulation gauges how well a heater can maintain a steady temperature. Heaters without thermostats were debited. Heat distribution shows how evenly these heaters warmed an unfurnished test room. Placement, room size, furnishings, and air leaks to other rooms or the outdoors affect performance. Spot heating: A good spot heater is appropriate for warming an individual office worker, say, or someone working in an unheated garage. Warmup time is a judgment of how much heat a heater produced five minutes after it was turned on.

❺ Noise. This judgment, important if a heater is used in a bedroom at night, covers the loudness of any fan present. The disadvantages note any intrusive tings and bonks a heater made as it cycled.

❻ Features. A heater with more than one wattage setting offers some versatility. If you need to warm a room only a little or warm a small room or maintain a heat setting in a warm room, a reduced wattage setting may do the job with less current draw. A thermostatic control is essential for maintaining a steady room temperature. A safety switch can be a tip-over switch (T) that shuts the heater off if it tilts too far or an overheat protector (O) that turns off the heater if it gets too hot; ceramic heaters have built-in overheat protection because their elements cannot get very hot.

Specifications and Features
All run on 120 volt AC current.

Except as noted, all: ● Have two-prong plug. ● Have on/off control separate from thermostat. ● Acceptable radiant models caused some scorching of fabric used in fire-safety test. ● Maintained surface temperatures below 150°F.

Key to Advantages
A—Has no cord; plugs directly into three-prong outlet.
B—Power-on or element-on light easy to see.
C—Well shielded, very difficult to stick fingers or foreign objects through grille.

D—Grille better than most at resisting probes from fingers or foreign objects.
E—Easy access for cleaning reflector, important for proper heating.
F—Easier than other upright types to wheel about.
G—Long power cord, almost 8 feet.
H—Tip-over switch trips when heater is tilted in any direction.
I—Has ground fault circuit interrupter to prevent electric shock. Safe for bathroom use.

Key to Disadvantages
a—Tip-over switch malfunctioned on CU's samples.
b—Recessed control knobs hard to operate.
c—Burn hazard—outside surfaces exceeded 200°F.
d—Burn hazard—grille exceeded 180°F.
e—Slight burn hazard—small part of grille exceeded 180°F.
f—No separate on/off switch; switched off with thermostat, an inconvenience.
g—Safety hazard—fingers could touch moving fan.

Key to Comments
A—Awkward to move from room to room or

up or down stairs.
B—Judged unlikely to tip, but failed tip-over test.
C—Fan can run without heat, but flow is much weaker than a portable room fan's.
D—Also sold as **Kensington DWP4K**, $170.
E—Has timer, but touchpads hard to press.
F—Floor temperature in front of grille—higher than most; may discolor flooring.
G—Has antifreeze setting that switches on heater if temperatures approach freezing.
H—Built-in humidifier had negligible effect.
I—Alarm sounds when tip-over switch activated.
J—Has cord storage.
K—Adjustable tilt.
L—Heater oscillates like an oscillating fan.
M—Mechanical timer can turn heater on and off.
N—Lighted clock can turn heater off, but not on.
O—Has lighted digital clock with bright display.
P—Has three-prong plug.
Q—Has two 750-watt sections (one for the radiator, one fan-forced convective) that can be run independently or together.
R—Comes with washable air filter.

Better ●　◒　○　◒　● Worse

Brand and model	Price	Type	Dimensions (HxWxD), inches	Weight, pounds	Stability	Overall safety	Temperature regulation	Heat distribution	Spot heating	Warmup time	Noise	No. wattage settings	Thermostat control	Safety switch	Advantages	Disadvantages	Comments
Medium-size and compact convection heaters																	
Holmes HFH-895	$98	Coil	17x18x6	8	○	○	◒	◒	○	◒	●	2	√	T	H	—	C,G,I,L,O
Arvin 29H40, A Best Buy	35	Coil	5x10x8	3	●	○	◒	○	○	◒	●	2	√	O	—		C,K
Arvin 29H85	53	Coil	13x8x6	5	○	○	◒	◒	◒	◒	◒	2	√	O	—		C,L
Patton HF-30W	80	Wire	15x10x6	5	○	○	○	◒	○	◒	○	3	√	O	B	—	C,E,G,M,N,R
Tatung EH1723	40	Coil	11x11x6	4	○	◒	◒	●	◒	◒	◒	3	√	O	—	—	C,J,L
DeLonghi TU02	40	Wire	5x10x7	4	●	○	◒	◒	○	○	◒	1	√	O	—	f	B,C,G,K
Holmes HFH-501	54	Coil	15x9x5	5	○	○	◒	◒	◒	◒	○	3	√	T	—	—	C,G,K,L
Patton 15S	55	Coil	9x11x6	4	○	○	○	◒	◒	◒	○	3	√	T	H	—	G,I
Sears 36224	64+	Coil	17x12x8	10	◒	○	●	◒	◒	◒	◒	2	√	T	B	—	C,M
Black & Decker HF230U	38	Coil	9x11x6	4	○	○	◒	◒	○	◒	◒	3	√	O	—	—	C
Lakewood 747-C	30	Coil	9x11x5	4	◒	◒	◒	◒	◒	◒	◒	2	√	—	—		B,C,G,R
Pelonis 1500W	120	Cer.	7x6x5	5	◒	◒	◒	○	○	○	○	①	√	O,T	D,G	e	E,C,G,O,Q
Patton BH-50	45	Wire	8x9x6	4	○	◒	②	○	○	◒	◒	3	—	O	I.	—	C
Arvin WH10	35	Coil	7x7x4	2	③	●	④	○	◒	○	○	1	—	O	A	o	C,R
Rival T-910	71	Coil	19x12x7	9	◒	○	◒	●	○	◒	◒	2	√	T	H	f,k	I

(continued)

	❶ Price	**❷** Type	**❷** Dimensions (HxWxD), inches	Weight, pounds	Stability	**❸ Safety** Overall safety	Temperature regulation	**❹ Performance** Heat distribution	Spot heating	Warmup time	**❺** Noise	**❻ Features** No. wattage settings	Thermostat control	Safety switch	Advantages	Disadvantages	Comments
Brand and model																	
Medium-size and compact convection heaters																	
Welbilt CH-100	90	Cer.	7x6x5	5	◐	◐	5	◐	◐	○	◐	1	—	O	C	e	C,R
Aladdin 43002	159	Cer.	7x6x5	6	○	◐	5	◐	○	◐	○	1	—	O	D,G	e	C,P,R
Tatung EH-5710	140	Cer.	7x6x5	5	◐	◐	5	◐	○	◐	○	1	—	O	D	e	C,R
Large convection heaters																	
Slant Fin AQ-1500	157	Water	11x72x4	23	◐	◐	○	○	●	◐	●	1	✓	T	G	g	A
Patton FL-40A	70	Calrod	10x40x7	8	◐	○	●	◐	●	◐	●	3	✓	—	B	—	B,G
Intertherm NP-1500-3	171	Water	9x74x4	24	◐	○	○	○	●	◐	●	3	✓	—	—	g	A
Heat Tech HT15	200	Steam	21x10x10	13	◐	○	●	◐	○	◐	◐	1	✓	T	B,H		J
Sears 36609	100+	Oil/B	9x47x4	13	◐	◐	○	◐	●	◐	●	2	✓	O	B	b	D
DeLonghi 3107T	80	Oil/Up	26x16x8	29	◐	◐	●	◐	◐	◐	●	3	✓	—	C	c	A,J,M
Welbilt RW-701P	70	Oil/Up	27x16x9	31	◐	◐	●	◐	◐	◐	●	3	✓	—	—	c	A,G,J
Delonghi F1-U	119	Oil/wire	23x14x7	20	◐	◐	◐	◐	◐	◐	◐	2	✓	—	F	c	G,J,Q
Radiant heaters																	
Arvin 49H50	80	Rib./B	9x40x5	11	◐	○	●	◐	○	○	◐	2	✓	T	B,D,E,H		I
Rival BB42-C	82	Rib./coil/B	8x38x5	10	◐	○	◐	◐	○	○	○	2	✓	T	B	f	—
Arvin 30H55	80	Rib.	13x19x5	8	○	◐	○	◐	○	○	●	2	✓	T,O	—	h	I,N
Lakewood 530	49	Rib.	11x19x7	8	○	◐	◐	◐	○	○	◐	2	✓	T,O	B	f	I
Toastmaster 2491	66	Rib.	13x18x6	8	○	◐	◐	○	○	○	○	1	✓	T,O	H	f	G,I
Not acceptable																	
■ *The following radiant heaters were judged not acceptable because of potential fire hazard. Listed alphabetically.*																	
Arvin 60H8106	65	Quartz/Up	29x12x13	6	◐	●	5	◐	◐	○	●	6	—	T	E	d	—
Presto 07892	73	Quartz/Up	28x7x9	6	○	●	2	◐	●	○	◐	3	—	T	E,H		I
Robeson 03-2701A	64	Quartz/H	15x19x7	6	○	●	2	◐	●	○	●	3	—	T	E,H	d	F,H

1 Wattage draw continuously adjustable.
2 No thermostat control, but manually adjusting heater to higher or lower wattage setting allows some control of room temperature.
3 Not applicable. Unit plugs directly into wall.
4 No means of controlling room temperature except by turning on or off manually.
5 No thermostat, but continuously variable control allows manual fine-tuning of room temperature.
6 No reduced wattage settings, but has control to cycle quartz tube on and off at varying rates.

Ratings of Light Bulbs

Listed by types. Within types, listed in order of overall quality based on measurements of life and light output. As published in a January 1990 report.

GUIDE TO THE RATINGS

❶ Product. Energy-saving bulbs run on 5 watts less power than the other bulbs in their group.

❷ Price. Average price paid. Bulb prices often are discounted, and rebates and coupons are common.

❸ Labeled lumens. The manufacturer's claim of light output, from the packages. Long-life bulbs sacrifice some light to gain extra life.

❹ Light output. An assessment of relative brightness when the bulbs were new and after 500 hours of testing (1,500 hours for the **Duro-Lites**). The tests included 36 bulbs per brand (18 for the **Duro-Lites**); to simulate home use, lights were turned off for 12 minutes every four hours. The tests showed that the manufacturers' estimates of lumen output are accurate, for the most part. The exceptions: the **Philips Econo-Miser** bulbs and the 60-watt **Duro-Lite,** which when new delivered 8 to 10 percent less light than promised. Light bulbs can be expected to dim somewhat over time, but some dim more than others.

❺ Life. What the manufacturer says on the label and an estimate of median life, in hours. The median is the number at which half the bulbs lasted longer and half burned out sooner. However, the lifetimes of individual bulbs can vary a lot.

6 % exceeded labeled life. The percentage of test samples that outlived their labeled lifetime. Another way of looking at longevity, this shows your chances of getting a long-lasting bulb, given the variability in the lifetimes of individual bulbs. (The relation of this number to the median lifetime depends on how tightly the lifetimes were clustered around the median.)

7 Cost per 1,000 hours. Typically, a bulb is used 1,000 hours a year. This figure was arrived at by dividing the cost of a bulb by its hours of life.

8 Efficiency. The average over the bulb's lifetime, measured in lumens per watt. The higher the number, the more efficient the bulb.

Better ⟵——————————⟶ Worse

1 Product	**2** Price	**3** Labeled lumens	**4** Light output	**5** Life, hours	**6** % exceeded labeled life	**7** Cost per 1,000 hours	**8** Efficiency, lumens per watt
60-watt bulbs							
Philips Soft White	$ .70	860	◖/◗	1000/1380	92%	$ .51	13.0
General Electric Miser	.89	855	◖/○	1000/ 984	44	.90	13.9
Sylvania Soft White	.70	870	◖/◗	1050/ 949	31	.74	13.1
General Electric Soft White	.77	855	◖/◗	1000/1018	53	.76	13.1
Sylvania Soft Energy Pincher	.87	870	◖/○	1000/ 855	17	1.02	13.8
Philips Econo-Miser	.95	860	◗/○	1000/1013	53	.94	13.4
Duro-Lite X2500 (long life)	1.70	780	◗/●	2500/4500+	83	.34	10.5
100-watt bulbs							
Philips Soft White	.67	1710	◖/○	750/ 820	72	.82	15.2
General Electric Soft White	.77	1710	◖/○	750/ 717	25	1.07	15.3
General Electric Miser	.90	1710	◖/◗	750/ 716	39	1.26	15.6
Sylvania Soft Energy Pincher	.93	1710	◖/◗	750/ 646	31	1.44	15.7
Sylvania Soft White	.70	1710	◖/◗	788/ 728	36	.96	14.5
Philips Econo-Miser	.90	1670	○/◗	750/ 856	75	1.05	14.7
Duro-Lite X2500 (long life)	1.61	1490	○/●	2500/2912	72	.55	13.1

Ratings of Low-Flow Shower Heads

Listed by type in order of preference as determined by a panel of users. Within type, bracketed models were about equally preferred, and are listed in order of increasing price. As published in a July 1990 report.

GUIDE TO THE RATINGS

1 Panelists' preference. These scores are derived from the experience of a use-test panel. Panelists were instructed to adjust each shower as desired, and afterward were asked to judge how well they liked it. The judgments are based on tests conducted at 80 psi pressure; the heads generally scored a little lower at 20 psi. (See Comments column for the models the panelists particularly liked or disliked at 20 psi.)

Shower heads that were liked most consistently earned the highest scores; those that were most disliked got the lowest. Those with middling scores, and those that were liked by some and disliked by others, were ranked in between.

2 Panelists' flow rate. This column lists the average flow rate, in gallons per minute, that the panelists actually used at the settings they chose.

3 Maximum flow rate. How many gallons per minute each shower head could produce with the supply taps fully open at low and high water-line pressures.

4 Spray and pulse settings. The major **spray** settings (**mist, fine, coarse**) and the

number of different **pulse** settings, if any, each shower head could produce. Some models, noted in the Comments, had numerous other settings as well. This column also notes the type of **adjustment** device each head provides—center (**C**), ring (**R**), or knob (**K**). Panelists least liked center adjustments, which force you to reach into the spray.

5 Materials/finish. The construction material and finish layer for each head: plastic (**P**), metal (**M**), or chrome-finished plastic (**C**). We noted no performance differences between a plastic and a metal head.

Specifications and Features
All: Attach to standard ½-inch diameter threaded shower pipe.

All hand-held models come with a 5-foot flexible hose, and, except as noted, a clip to hang them from shower inlet pipe.

Key to Comments
A—Consistently liked at 20 psi.
B—Consistently disliked at 20 psi.
C—Center-knob adjustment; judged less desirable than other types.
D—Judged noisier than most at 80 psi.
E—Judged too feeble to use at 20 psi.
F—Head attaches to wall mount (included).
G—Has shut-off valve.
H—Side-mounted adjustment lever.
I—Can mix spray with pulse.
J—Side-mounted adjustment knob.
K—Can mix fine and coarse spray.
L—Force continuously adjustable on each setting.
M—Comes with extension pipe.

Better ● ◖ ○ ◗ ● Worse

Brand & model	Price	Panelists' preference ①	Panelists' flow rate, gpm ② At 20 psi	At 80 psi	Maximum flow rate, gpm ③ At 20 psi	At 80 psi	Mist	Fine	Coarse	Pulse (no.)	Adjustment	Materials/finish	Comments
Fixed-position shower heads													
Sears Energy-Saving Shower Head 20170	$ 6	◖	2.1	2.6	2.4	2.9	—	✓	—	—	—	M/C	—
Thermo Saver DynaJet CF01	6	◖	1.1	2.3	1.3	2.8	✓	—	—	—	—	M	G
Resources Conservation The Incredible Head ES-181	7	◖	1.0	1.9	1.0	2.0	✓	—	—	—	—	M	B,G
Zin-Plas Brass Showerhead 14-9601-F	8	◖	1.5	2.9	1.9	3.6	—	✓	✓	—	K	M	A,H
Zin-Plas Water Pincher 14-9550	10	◖	1.2	2.4	1.4	2.7	—	✓	—	—	—	M	A
Whedon Saver Shower SS2C	10	◖	1.1	2.1	1.3	2.5	✓	—	✓	—	—	M	G
Great Vibrations Water Saver Massage B28400	11	◖	1.0	2.4	1.2	2.6	—	✓	✓	—	C	P	C,G
American Standard Shower Head Chrome 10509.0020A	15	◖	.9	2.5	1.3	2.9	—	✓	✓	—	C	M/C	C
Teledyne Water Pik Shower Massage 5 SM-2U	24	◖	1.9	2.4	2.4	2.7	—	✓	—	2	R	M/C	I
Chatham Solid Brass Shower Head 44-3S	39	◖	1.8	2.7	2.0	2.8	—	✓	—	—	K	M	J
Teledyne Water Pik Shower Massage 8 SM-4	40	◖	1.1	2.3	1.3	2.5	—	✓	✓	4	R	M/C	K
Melard Water-saving Adjustable 3610	10	○	1.2	2.4	1.4	2.8	—	✓	✓	—	K	C	B,G,J
Pollenex Dial Massage DM150	15	○	1.9	2.0	2.3	2.2	✓	✓	—	2	R	P	D,L
Nova B6402	19	○	1.1	2.0	1.2	2.5	—	✓	—	—	—	M	G
Speakman Anystream S2253-AF	29	○	1.6	2.3	2.0	2.5	—	✓	✓	—	K	M/C	J
Kohler City Club K-7351	75	○	1.8	2.6	2.2	2.8	—	✓	✓	—	K	M	B,J
NY-Del 550-II	3	◗	—	1.7	1.1	1.8	—	✓	✓	—	C	P	C,E
Ondine Water Saver 28446	9	◗	1.7	2.4	2.3	2.7	—	✓	✓	—	C	M/C	C
Kohler Trend 11740	10	◗	1.3	2.4	2.1	2.7	—	✓	✓	—	C	M/C	C
Alsons Somerset 673	11	◗	1.8	3.2	1.8	4.7	—	✓	✓	—	C	M/P	C
Speakman Cosmopolitan S2270-AF	12	◗	1.9	2.2	2.3	2.8	—	✓	✓	—	C	M/P	C
Pollenex Dial Massage DM109	17	◗	1.2	1.5	1.3	2.7	—	✓	✓	2	R	P	B,D,L
Alsons Alspray Massage Action 690C	33	◗	1.2	2.4	1.8	3.1	—	✓	—	1	K	C	B,J
Moen Pulsation 3935	45	◗	1.6	2.0	1.8	2.9	—	✓	—	1	K	M/P	J,M
Hand-held shower heads													
Sears Personal Hand Shower 20173	23	●	1.5	3.1	1.6	3.4	—	✓	—	1	R	M/C	—
Teledyne Water Pik Shower Massage 5 SM-3U	43	●	1.2	2.3	1.2	2.6	—	✓	—	2	R	M/P	I
Alsons Hand Shower 462PB	11	◖	1.1	2.5	1.3	2.6	—	✓	—	—	—	P	G
Alsons Massage Action Pulsating 45C	58	◖	1.5	2.5	1.6	2.7	—	✓	—	1	R	M/P	—
Moen Pulsation 3981	95	◖	1.3	2.2	1.3	2.4	—	✓	—	1	K	M/P	F
Teledyne Water Pik Super Saver SS-3	16	○	1.1	2.2	1.2	2.5	—	✓	—	—	R	M/P	—
Pollenex Dial Massage DM209	24	○	1.8	2.5	2.2	5.4	—	✓	✓	2	R	M/P	A,L
Pollenex Dial Massage/Steamy Mist DM230	29	○	1.5	3.1	2.3	4.9	✓	✓	—	2	R	M/P	L
Pryde Splash 2461	35	○	1.1	2.6	1.7	2.8	—	✓	—	1	R	P/C	B
Teledyne Water Pik Shower Massage 8 SM-5	44	○	1.6	2.1	1.9	2.5	—	✓	✓	4	R	P/C	K

Listed in order of estimated quality. Bold rules separate distinct differences in overall performance. Brackets indicate essentially identical overall performance. As published in a November 1989 report.

GUIDE TO THE RATINGS

❶ Price. All prices include automatic ice-maker, either as factory installed or as kit (with a kit, you may have to pay for installation labor).

❷ Annual energy use. These costs, based on Consumers Union's energy-consumption tests, differ somewhat from figures on the "energy guide" labels in the stores. CU's test was more stringent than the government requires for that label. The estimates are based on a 37°F refrigerator and a 0° freezer and assume that energy-saver (anticondensate) heaters are on for six months and off for six. Energy use does not include operation of the ice-maker.

To determine what the annual dollar cost of operation in your area might be, multiply your electric rate (in cents per kilowatt-hour) by the number of kilowatt-hours used by the model and divide by 100.

❸ Temperature balance. The ability of a model to maintain 37°F in the refrigerator and 0° in the freezer. Only the **Ward's** could not achieve that ideal balance.

❹ Temperature uniformity. The degree to which temperatures varied throughout refrigerator and freezer main space and doors.

❺ Temperature compensation. How well did each model compensate for changes in room temperature? Tests simulated a winter vacation, when you might turn the thermostat down from 70° to 55°F, a hot summer day when your kitchen might go from 70° to 90°F, or a fall day when temperatures might drop from 90° to 70°F.

❻ Reserve capacity. How well each model worked under the most severe conditions—a possible indication of how it will perform after components have aged.

❼ Meat-keeper. Meat and cold cuts are best kept at temperatures between 30° and 35°F. Some meat-keepers were actually warmer than the center of the refrigerator.

❽ Crisper humidity. How well this area retained moisture—important if you don't want lettuce and such to dry out.

❾ Ice-maker. Scores reflect the amount (in pounds) of ice made per day, frequency of ice-making cycles, and size of storage bin.

❿ Noise. The opinion of a listening panel, plus measurements with a sound-level meter, as the compressor came on and while it ran.

⓫ Overall convenience. Each model was subjected to a 70-point convenience inspection. How clear and logical were the controls? How easy was it to clean? Unique features are listed in the Advantages (or Disadvantages) column.

⓬ Dimensions. Height, width, and depth to next highest quarter inch. Door hinges and handles are included.

⓭ Depth, door open. Use this figure to determine clearance needed in front of refrigerator to open the door 90 degrees.

⓮ Capacity. A measurement of usable storage space. Manufacturers'-claimed capacities, which are certified by AHAM, are 25 to 35 percent higher. Consumers Union did not include shelf retainers and nooks and crannies. Total volume and shelf area are without (and with) automatic ice-maker.

Specifications and Features

All have: ● Adjustable glass shelves in the refrigerator main space. ● Reversible, textured steel doors. ● Four rollers. ● Leveling provisions. ● Anticondensate, "energy-saver" switch. ● Door stops.

Except as noted, all have: Controls to maintain 37°F in refrigerator and 0°F in freezer. ● Plastic seamless liner. ● At least one 40-watt light in refrigerator. ● One-year warranty on parts and labor for entire unit. ● Five-year warranty on sealed refrigeration unit. ● Two crispers, one or both with seals.

Key to Advantages

A—Energy-saver switch has light or red bar to remind you that anticondensate heater is on.
B—Controls easy to reach.
C—Meat-keeper has temperature control; judged moderately effective.
D—Has porcelain-on-steel liner, meat-keeper, and crispers; judged easy to use and clean.
E—Adjustable snuggers or fingers in refrigerator or freezer door shelves keep small items upright.
F—Some refrigerator door shelves or door-shelf bins are removable and adjustable.
G—Some refrigerator or freezer door-shelf guards are removable for easy cleaning.
H—Has refrigerator shelf that can also serve as a can dispenser or defrost area.
I—Has at least one deep door shelf for gallon container.

J—Has extra-deep freezer door shelves.
K—Has freezer door area for frozen cans.
L—Has ice-cube tray storage shelf if ice-maker is not installed.
M—Has light in freezer.
N—Freezer vents judged less vulnerable to spills than others.
O—Has rear condenser coils; needs less cleaning than bottom-mounted coils.
P—Freezer shelf is adjustable for more than two positions.
Q—Has locks on refrigerator wheels.
R—One crisper is larger than other, offering more flexibility.
S—Some drawers mounted on rollers for easier handling.
T—Anticondensate heater very effective.

Key to Disadvantages

a—Freezer temperature could not be made cold enough.
b—Refrigerator temperature varied much more than average under stable conditions.
c—Freezer warmed more than most during defrost.
d—Exterior sweated more than others with anticondensate heaters on in moderate humidity.
e—Markings on energy switch could be confusing.
f—Butter door will not stay open by itself.
g—Lacks butter tray.
h—Freezer shelf not adjustable.
i—Drip pan cannot be removed and therefore is very hard to clean.
j—Condenser judged very hard to clean.
k—Requires a minimum 3-inch clearance on top, 1 inch on bottom.
l—Optional ice-maker very difficult to install.
m—Drip pan has to be lifted up over wire brace and could spill.
n—Handle had to be installed with a Torx-head screwdriver.
o—Crisper drawers lack adequate stops.
p—Crisper shelf-support can be dislodged easily when opening crisper drawers.
q—Door-shelf retainer lets 2-liter soda bottle topple out when door is abruptly opened.
r—Gap under refrigerator door-shelf retainer lets small items fall through.
s—Gap under freezer door-shelf retainer lets small items fall through.
t—Position of energy-saver switch is not immediately obvious

Key to Comments

A—Neither crisper has seals.
B—Meat-keeper has seals.
C—Light bulb is shielded well.

(continued)

Brand and model	Price (and range) [1]	Cost	Annual energy use Kilowatt-hours [2]	Balance [3]	Uniformity [4]	Temperature Compensation [5]	Reserve capacity [6]	Meat-keeper [7]	Crisper humidity [8]	Ice-making [9]	Noise [10]	Overall convenience [11]
Amana TC20M	$865	$ 79	1026	◉	◒	◉	◉	◉	○	◒	○	◒
Sears Kenmore 79061	776	84	1094	◉	○	◒	◒	◒	◒	○	◒	◒
General Electric TBX21ZL	818	84	1089	◉	○	◉	○	◉	◒	◒	○	○
Maytag RTD21A	949	89	1154	◉	○	◒	○	◉	◒	◉	◒	◉
Whirlpool ET20DKXV	795	84	1094	◉	○	◒	◒	◒	○	○	◒	◒
Jenn-Air JRT1214	842	89	1154	◉	◒	◉	○	◒	◉	◒	○	◒
Caloric GFS207-1	740	84	1094	◉	○	◒	◒	◒	◒	◒	○	◒
Hotpoint CTX21GL	723	84	1089	◉	◒	◉	○	◉	●	◒	○	◒
Frigidaire FPCE21TFW0	806	86	1116	◉	○	○	○	○	○	○	◒	○
Montgomery Ward HMG21974	759	89	1154	○	◒	◒	◒	◒	◉	◒	○	◒
Gibson RT21F9WT3B	825	83	1077	◉	●	◒	◒	●	◒	○	◒	◒
KitchenAid KTRF20KT	856	91	1182	◉	○	◒	◒	○	●	○	◒	◒
Tappan 95-2187-00-5	764	86	1116	◉	○	○	○	○	○	◒	◒	◒
Kelvinator TMK206ENOW	631	83	1077	◉	●	◒	◒	●	◒	○	◒	◒
Magic Chef RB21JN-4A	797	101	1308	◉	◒	◒	●	◉	◒	○	○	○
Admiral NT21K9	794	101	1308	◉	◒	◒	●	◉	◒	○	○	○

◉ ◒ ○ ◓ ●
Better ⟵——————⟶ Worse

D—Porcelain-on-steel liner has 10-year warranty.
E—"Picture-frame" door trim.
F—Door trim accepts optional door panel.
G—White handles.
H—Brown door-seal gaskets.
I—Almond-colored door interior.
J—Freezer sticker describes food storage times.
K—60-watt bulb in refrigerator.
L—1-year warranty for parts (not labor).
M—10-year warranty on compressor.
N—Separate wine-bottle holder.
O—Vertical divider in freezer.
P—Two lights in refrigerator.
Q—Lacks egg-storage provision.
R—Leveling is done by adjusting rollers.
S—Leveling is done by separate leveling legs; that prevents rolling.
T—According to the manufacturer, model has been replaced by an essentially similar model.
U—Although the model has been discontinued and no longer is available, the information has been retained to permit comparisons.

Dimensions (HxWxD), inches	Depth, door open, inches	Capacity, cubic feet			Shelf area, square feet	Advantages	Disadvantages	Comments
		Refrigerator	Freezer	Total				
68x32¼x32¼	61	10.1	4.6	14.7 (13.7)	25.2 (24.0)	C,E,F,H,L	c	G,N,O,R
66¼x32¾x30¼	59¾	11.1	4.1	(15.2)	(23.1)	A,B,E,F,I,J,M	h,m,n	P,R,U
66¼x31¼x31¾	60	10.0	5.4	15.4 (14.6)	23.7 (22.9)	C,E,G,J,L,P,R	g,i,q	B,G,H,I,R
66¼x31½x33¼	61¼	10.5	5.4	15.9 (14.7)	28.1 (26.8)	A,B,C,E,F,J,L,M,P,Q,S	c,j,q,s	B,C,H,P,R
66¼x32¾x30¼	59¾	10.6	4.7	15.3 (14.4)	26.5 (25.5)	A,B,E,F,I,L	g,h,m	H,P,Q,R,U
66½x31½x31¾	61½	11.2	4.1	(15.3)	(25.1)	B,C,E,F,J,T	e,j,o,p,t	F,S,U
66¼x32¾x30¼	59¾	10.5	4.9	15.4 (14.8)	25.1 (24.3)	A,B,E	g,m,q,r	A,L,P,R
66¼x31¼x31¾	60	10.9	5.5	16.4 (15.6)	24.2 (23.5)	C,G,J,L	g,i,o,p,q	A,B,R
66¾x31¼x30½	59¾	10.4	5.1	15.5 (14.9)	26.7 (25.9)	C,E,F,J,L,O,R	d,h,i,k,l,o,p,s	H,I,S,T
66½x31½x31¾	61½	11.3	5.4	16.7 (15.8)	27.7 (26.7)	B,C,E,J,T	a,j,o,p,r,t	I,J,S,U
66¾x31¼x31½	60	10.9	5.0	15.9 (15.3)	23.4 (22.6)	E,F,G,H,K,O,R	b,d,g,i,k,l,o,p,r	E,I,K,M,Q,S,U
66¼x32¾x29¼	58¾	9.6	4.6	14.2 (13.5)	22.9 (22.0)	A,D,J,M,N,S	f,h,m,o,r,s	B,C,D,R,T
66¾x31¼x30¾	59¼	11.1	5.1	16.2 (15.6)	23.6 (22.9)	G,O,R	d,f,g,h,i,k,l,o,p,q	A,I,Q,S,U
66¾x31¼x30¾	59¼	10.9	5.0	15.9 (15.3)	23.3 (22.6)	G,O,R	b,d,f,g,i,k,l,o,p,q,r	F,K,S
66¼x31¾x31¾	61	10.1	5.6	15.7 (14.6)	26.4 (25.3)	B,C,E,F,J	e,j,o,q	A,G,S
66¼x31¾x32½	61	10.3	5.5	15.8 (14.9)	26.7 (25.7)	B,C,E,J	e,j,o,q	A,G,S,T

Ratings of Side-by-Side Refrigerator/Freezers

Listed in order of estimated quality. Ratings should be used with Features table. Bracketed models had essentially identical overall performance and are listed alphabetically. The *Amanas* are essentially the same except for controls and defrost systems. As published in a May 1991 report.

GUIDE TO THE RATINGS

❶ **Brand and model.** All are large, top-of-the-line models with a factory-installed automatic ice-maker and a dispenser in the door for ice and water.

❷ **Price.** Does not include plumbing hardware for connecting ice-makers or water dispensers.

❸ **Annual energy use.** These figures are based on energy-consumption tests with refrigerators set for 37°F and freezers set for 0°F and assuming any anticondensate heater ("energy saver") is on for six months. The temperature criteria are more stringent than those specified in Government test protocols. Figures do not include the operation of ice-maker or ice and water dispensers. Costs are based on the 1990 national average electricity rate of 7.9 cents per kilowatt-hour. To determine the cost in your area, multiply your electric rate (cents per kWh) by the number of kilowatt-hours the model uses per year.

❹ **Temperature balance.** A model's ability to maintain the set temperatures: 37° in the refrigerator and 0° in the freezer.

❺ **Temperature uniformity.** The degree to which temperatures varied throughout the refrigerator and freezer main space and doors. Normally, the dairy compartment in the door runs several degrees warmer and

(continued)

① Brand and model	② Price	③ Annual cost	③ Kilowatt-hours	④ Balance	⑤ Uniformity	⑥ Compensation	⑦ Reserve capacity	⑧ Meat-keeper	⑨ Crisper humidity	⑩ Ice & water dispensers	⑪ Ice-making	⑫ Noise	⑬ Overall convenience
						Energy use				Temperature			
Amana SZD27K & SZDE27K	$1490/1780	$116	1464	◉	◐	◉	◐	◉	◉	○	◐	○	◐
General Electric TFX27VL	1475	115	1460	◉	○	○	○	◉	◐	◉	◉	○	◐
Hotpoint CSX27DL	1320	115	1460	◉	○	○	○	◉	◐	◉	◉	○	◐
Sears Kenmore 50771	1385	115	1460	◉	○	○	○	◉	◐	◉	◉	○	◐
RCA MSX27XL	1490	123	1560	◉	○	○	○	◉	◐	◉	◉	◐	◐
Whirlpool ED27DQXW	1520	124	1566	◐	◐	◐	◐	◉	◐	◐	◐	○	◐
KitchenAid KSRS25QW	1535	114	1440	◉	○	○	◐	◐	◐	◐	◐	○	◐
General Electric TFX27FL	1685	133	1680	◉	○	○	◐	◉	◐	◉	◐	○	◐
Admiral CDNS24V9	1065	109	1380	◉	◐	○	○	○	●	◐	◐	○	○
Jenn-Air JRSD246	1355	107	1356	◉	◐	◐	◐	◉	●	◐	◐	◐	◐
Maytag RSW24A	1580	107	1356	◐[1]	◐[1]	◐	◐	○[1]	●	◐	◐	○	◐
Montgomery Ward (Signature 2000) 24902	1220	125	1584	◉	○	○	○	◐	●	○	◐	◐	○
Frigidaire FPCE24VWL	1100	116	1472	◉	◐	○	○	◐	◐	◐	◐	○	○
Kelvinator FMW240EN	1235	116	1472	◉[2]	◐	○	○	◐	◐	○	◐	○	○
White-Westinghouse RS249M	1000	116	1472	◉	◐	○	○	◐	○	◐	○	◐	○

[1] Repositioning meat-keeper and utility drawer may improve or worsen temperature performance.

[2] Freezer on two samples could not reach zero in some tests because absence of baskets allows inlet air ducts to be blocked.

the meat-keeper several degrees cooler than the main space.

⑥ **Temperature compensation.** How well each model could compensate for changes in room temperature. Tests simulated a winter vacation, with the household thermostat turned down from 70° to 55°; a hot summer day, when a kitchen might go from 70° to 90°; and a fall day when temperatures might go from 90° to 70°.

⑦ **Reserve capacity.** How well each model worked under severe conditions, a possible indicator of how it will perform after components have aged.

⑧ **Meat-keeper temperature.** Fresh meat and coldcuts are best kept at temperatures between 30° and 35°.

⑨ **Crisper humidity.** These judgments show ability to retain moisture and effectiveness of humidity control, if any.

⑩ **Ice and water dispensers.** Scores reflect the coolness of five successive 10-ounce tumblers of water and how well ice was dispensed into glasses.

⑪ **Ice-making.** Scores reflect the quantity of ice produced per day, which ranged from 3½ to 5 pounds. The frequency of ice-making cycles and the size of the ice bin were also taken into account.

⑫ **Noise.** Scores reflect listeners' judgments and readings taken with a sound-level meter.

⑬ **Overall convenience.** Each model was inspected with a checklist, covering details about controls, door handles, lighting, layout of shelves, and much more. Especially important: how easy the controls were to read and set and how easy the refrigerator was to clean.

⑭ **Dimensions.** The measurements are to the next highest quarter-inch, door hinges and handles included. Depth, with the refrigerator door open at a 90-degree angle, measured from 49 to 50½ inches.

⑮ **Side clearances.** How much space is needed between the hinge side of an open door and an adjacent right-angled wall to remove a crisper, say, or pull out shelves and

bins. First figure is for main compartment, second for freezer door.

⑯ **Capacity.** The usable internal volume, to the nearest tenth of a cubic foot. The figures are smaller than manufacturers' (see Features table) because they do not include shelf retainers, crannies that never get used, and other unusable space such as room taken by the ice-maker.

⑰ **Shelf area.** Again, the measurements are more realistic than those of manufacturers.

Features in Common

All have: ● Adjustable glass shelves in refrigerator. ● Temperature-controlled meat-keeper. ● Anticondensate heating, judged effective in controlling exterior condensation. ● At least one crisper drawer, at least partly sealed. ● Bulk-storage drawer in bottom of freezer. ● Door-shelf retainers, bins, or both, that are easily removed for cleaning. ● Four rollers on base, the front pair adjustable for leveling.

Except as noted, all: ● Can maintain 37°F in

Dimensions (H×W×D), in.	Side clearances, in.	Capacity, cu. ft.	CU-measured shelf area, sq. ft.	Advantages	Disadvantages	Comments
68½x35¾x35¾	5/12	16.7	24.8	C,D,E,G,H,I,J,O,P,R,S,W,Y	—	C,D,G,I,J,M,O
69¾x36x32¾	12/11	16.7	23.1	C,E,G,H,K,O,U,W	a,h	B,E,F,I,N
69¾x36x32¾	12/11	17.2	22.0	C,E,G,H,K,U,W	a,h	B,N
69¾x36x32¾	12/11	16.7	21.5	A,C,F,G,H,O,U,W	a,h	B,I,J,N,R
69¾x36x32¾	12/11	16.7	23.1	C,E,G,H,K,O,U,W,X	a	B,N
69½x35½x34	10½/11	15.7	22.0	A,B,C,D,F,G,H,I,K,M,W	j	B,E,G,I,J,K,O
69½x35½x32¼	12/11	16.0	24.3	A,C,F,G,I,L,M,W	j	B,D,E,G,I,J,K,O
69¾x36x33¼	12/11	16.4	23.0	C,E,G,H,K,O,U,W,X	a,h	B,E,F,N
67x36x31¼	17/11	16.0	22.5	C,H,K	f,i,l	B,D,H,I,L
67x36x30¾	17/11	15.3	23.3	C,E,H,I,L	f,i	B,C,G,H,I,J,L
67x36x31½	17/11	15.1	25.0	C,E,H,I,K,L,O,Q,V	d,i	B,E,G,H,I,J
67x36x30¾	17/11	16.1	23.7	H,K,M,Q	f,i,k,l	A,B,E,H,I,K,L,Q
66¾x36x32	6½/9	15.4	22.0	C,F,H,I,N,O,T	b,e,f,g,m	E,G,I
66¾x36x32½	6½/9	16.0	22.7	T	b,c,e,f,m	E
66¾x36x32¼	6½/9	15.4	24.2	F,K,T	b,c,e,f,m	H,I

refrigerator, 0°F in freezer. ● Have adjustable shelves/bins on refrigerator door. ● Have textured steel doors. ● Have plastic seamless liner. ● Have door stops. ● Met 1990 U.S. Energy Dept. efficiency requirements.

Key to Advantages
A—Controls are easy to reach.
B—Some refrigerator shelves slide forward for easier access.
C—Refrigerator door has removable bins whose height can be adjusted.
D—Freezer door has removable bins whose height can be adjusted.
E—Height of some freezer shelves adjustable.
F—Freezer baskets slide for easier access. (On **White-Westinghouse** and **Frigidaire,** baskets can be removed to convert space to shelves.)
G—Refrigerator door has at least one shelf that can easily fit one-gallon milk containers.
H—Bookend-type "snugger(s)" on refrigerator door shelves keep small containers upright (**Frigidaire** uses plastic fingerlike tabs).
I—Two crisper drawers.
J—Sliding freezer shelf for cans.

K—Separate deli/utility drawer.
L—Some drawers are on rollers.
M—Fast-freeze shelf above ice-maker.
N—Juice-dispenser jug on inside of door.
O—Comes with microwaveable containers for storing, heating, and serving.
P—Beverage area on door shelf has separate temperature control.
Q—Separate ice and water dispenser lock-outs.
R—**SZDE27K** has electronic touchpad controls and electronic monitor.
S—Crisper's humidity controls very effective.
T—Has rear condenser coils; need cleaning less often than bottom-mounted coils but refrigerator must be moved from wall to clean.
U—Defrost drain-tube opening easily accessible for cleaning.
V—Locks on rollers to curb rolling.
W—Lets user select cubes or crushed.
X—Electronic monitor and diagnostic system.
Y—Dispenser area has night light, which can he set to turn on automatically after dark.

Key to Disadvantages
a—Upper shelf area of freezer door warmed more than most during defrost.

b—Lacks separate On/off control arm for ice-maker; ice bin must be physically removed to stop ice-maker.
c—Butter door will not stay open by itself.
d—Ice and water dispensers difficult to operate.
e—Requires at least three-inch clearance on top and one inch on bottom for air circulation.
f—Lacks door stops.
g—Doors are smooth, not textured; may show smudges more readily.
h—Drip pan cannot be removed; hard to clean (but water may evaporate faster).
i—Condenser wrapped in metal "shroud," which curbs dust but makes cleaning hard.
j—Drip pan must be lifted over a wire brace; hard to remove.
k—Lacks adjustable shelves on refrigerator door.
l—Crisper drawers lack adequate stops.
m—Ice dispenser can be activated even when freezer door is open; ice can fall to floor.

Key to Comments
A—Tested model manufactured in 1989 (1990 version not available at time of tests). Did not meet 1990 Government energy-efficiency requirements.
B—Meat-keeper is sealed.
C—Standard door trim accepts optional decorator door panels.
D—Has white handles.
E—Has brown (or tan) door-seal gasket.
F—Has almond-colored door interior.
G—Crisper(s) have humidity control.
H—Chart in freezer lists food-storage times.
I—Has wine rack.
J—Has two lights in refrigerator.
K—Has two lights in freezer.
L—Has separate leveling legs to prevent rolling.
M—Entire cabinet, not just doors, is textured.
N—Has metal interior liner.
O—Unlike other models that circulate hot refrigerants, these models have an electric "energy saver" anticondensate heater, judged effective in controlling exterior condensation. (**Whirlpool** and **KitchenAid** have light to remind you that heater is on.)
P—Has door-within-door Refreshment Center.
Q—Although discontinued, the information has been retained to permit comparisons.
R—Model has been discontinued and replaced by **51771** ($1530 suggested retail), which company says is essentially similar.

Features of Side-by-Side Refrigerator/Freezers

❶ Brand and model. Models tested and a selection of other models of similar design and capacity. Brands are listed alphabetically; within brands, in order of increasing price. Except for tested models (marked with ●), information is based on an interpretation of literature from manufacturer and from the Association of Home Appliance Manufacturers.

❷ Price. Retail price, or suggested retail as quoted by the manufacturer.

❸ Energy cost. From a model's Energy-Guide sticker, this is an annual figure, based on U.S. Department of Energy test protocols and the 1988 average electricity rate of 8.04 cents per kilowatt-hour. (Energy costs in the Ratings are based on test protocols and the 1990 rate of 7.9 cents per kWh.)

❹ Claimed capacity. Volume as given by the manufacturer. Capacity listed in the Ratings is more realistic, however.

❺ Claimed shelf area. The total claimed area. Again, the Ratings give a more realistic figure.

❻ Shelves. The number and type of shelves in each compartment. A **p** indicates a partial-width shelf; all others are full-width. The number of drawers (or baskets) in the freezer is noted too.

Key to Comments

A—Door-trim frame and/or optional door panels are available.

B—Optional ice-cream maker available.

C—Ice dispenser delivers crushed ice or cubes (selectable).

D—Has separate snack drawer.

E—Has wine rack.

F—Has separate storage dishes.

G—Has Refreshment Center door within refrigerator door.

H—Freezer shelf is adjustable.

I—Has dispenser jug on refrigerator door shelf.

J—Has "bookend" or other type of shelf snugger to hold small containers on refrigerator door shelf.

K—Some refrigerator door shelves/bins are adjustable.

L—Some freezer door shelves are adjustable.

M—Shelves slide out for easy access.

N—Freezer has quick-freeze shelf.

O—Freezer has shelf for juice cans.

P—Refrigerator door shelf not deep enough to hold gallon container comfortably.

Q—Refrigerator's main space also has one plastic shelf.

R—Although discontinued, the information has been retained to permit comparisons.

S—Model has been discontinued and replaced: **Ward 24894** by **24824**, $900; **Sears 50771** by **51771**, $1530; **Sears 51791** by **51781**, $1729.

❶ Brand and model	❷ Price	❸ Energy cost, cu. ft.	❹ Claimed capacity Refrigerator, cu. ft.	Freezer, cu. ft.	Total, cu. ft.	❺ Claimed shelf area, sq. ft.	Crisper humidity control	Ice & water dispensers
Admiral								
● CDNS24V9	$1199	$105	15.2	8.2	23.5	27.6	—	✓
BDNS24-9	1199	105	15.1	8.4	23.6	28.1	—	✓
Amana								
● SZD27K	1700	105	16.9	9.8	26.7	28.3	✓	✓
● SZDE27K	1900	105	16.9	9.8	26.7	28.3	✓	✓
Frigidaire								
● FPCE24VWL	999	106	14.7	9.3	24.0	26.3	✓	✓
FPCE24VF	999	90	14.7	9.3	24.0	29.9	✓	—
GE								
TFX27RL	1550	113	16.7	9.9	26.6	28.8	—	✓
● TFX27VL	1600	106	16.7	9.9	26.6	29.9	—	✓
TFX27EL	1650	121	16.7	9.9	26.6	29.9	—	✓
TFX27FL	1850	125	16.7	9.9	26.6	29.8	—	✓
Hotpoint								
● CSX27DL	1400	106	16.7	9.9	26.6	28.8	—	✓
CSX27CL	1650	125	16.7	9.9	26.6	29.8	—	✓
Jenn-Air								
● JRSD246	1750	105	15.3	8.2	23.5	27.5	✓	✓
Kelvinator								
● FMW240EN	899	106	14.7	9.3	24.0	26.4	—	✓
KitchenAid								
● KSRS25QW	1649	109	15.2	9.8	25.0	27.2	✓	✓
Maytag								
RSD24A	1430	90	15.2	8.6	23.8	30.4	✓	—
● RSW24A	1700	100	15.2	8.3	23.5	29.0	✓	✓
Montgomery Ward								
24894	—	102	15.1	8.7	23.8	28.7	—	—
● 24902	1200	123	15.1	8.5	23.6	28.1	—	✓
RCA								
● MSX27XL	1550	121	16.7	9.9	26.6	29.9	—	✓
Sears Kenmore								
● 50771	1385	113	16.6	9.9	26.5	27.6	—	✓
51791	—	125	16.6	9.9	26.5	27.5	—	✓
Whirlpool								
● ED27DQXW	1625	113	16.4	10.2	26.7	25.6	✓	✓
White-Westinghouse								
● RS249M	949	106	14.7	9.3	24.0	29.5	—	✓

Better ← → Worse

Electronic monitor or controls	Gal.-container door rack	⑥ Refrigerator shelves			⑥ Freezer shelves					Comments
		Main space, glass	Door, aluminum	Door, plastic	Main space, glass	Main space, wire	Main space, plastic	Drawers	Door, plastic	
—	√	4	—	5	3	—	1p	1	5	D,E,J,K,L,P
—	√	4	—	5	3	—	1+1p	1	5	A,B,D,E,J,P
—	√	3+2p	—	4+3p	—	3	—	1	6	A,C,E,F,H,K,L,O
√	√	3+2p	—	4+3p	—	3	—	1	6	A,C,E,F,H,K,L,O
—	—	4	3	—	—	1	2p	3	5	A,E,F,I,K
—	—	4	3	—	—	5	—	1	6	A,E,F,I,K
—	√	4	—	5	—	3	—	1	5	A,C,D,F,H,K
—	√	4	—	5	—	3	—	1	5	C,D,E,F,H,J,K
√	√	4	—	5	—	3	—	1	5	A,C,D,F,H,K
√	√	4	—	5	—	3	—	1	5	A,C,D,F,G,H,J,K
—	√	4	—	5	—	3	—	1	5	A,C,D,H,J,K
√	√	4	—	5	—	3	—	1	5	A,C,D,F,G,H,J,K
—	√	4	—	5	—	4	—	1	5	A,E,H,J,K,P
—	—	4	5	—	—	3	2p	1	5	A,K,Q
—	√	5	—	4	—	1	1	3	5	A,C,E,K,N
—	√	4	—	5	—	4+1p	—	1	6	A,D,E,F,H,J,P
—	√	4	—	5	—	4	—	1	5	A,D,E,F,H,J,P
—	√	4	—	5	5	—	—	1	6	D,E,J,N,P,S
—	√	4	—	5	3	—	1p	1	5	D,E,J,N,P,R
√	√	4	—	5	—	3	—	1	5	C,D,F,H,J,K,L
—	√	4	—	5	—	—	3	1	5	A,C,E,F,J,K,M,S
—	√	4	—	5	—	—	3	1	5	A,C,E,F,G,J,K,M,S
—	√	4	—	4	—	—	1	4	5	C,D,E,J,K,L,M,N
—	—	4	4	—	—	2	2p	2	5	A,D,E,K,Q

Ratings of Clothes Washers

Listed in order of estimated quality. Differences between closely ranked models were slight. All should wash well. As published in a February 1991 report.

GUIDE TO THE RATINGS

1 Brand and model. The Ratings include popular, low-priced, two-speed models.

2 Price. Estimated average price.

3 Dimensions. Height includes control panel on top of machine; work surface of all models can be set to 36 inches with leveling legs (extendable to 1½ inches). Parenthetical figure is height with lid open or depth with door open. The **Amana, Maytag,** and **Speed Queen** can be installed flush to wall; others need up to 4 inches extra in back to allow room for hoses.

4 Lid opens. Check your installation site to see which direction is most convenient for you.

5 Load capacity. Based on the largest load a machine could handle and still provide adequate wash action.

6 Water efficiency. Based on gallons used per amount of laundry accommodated (at maximum fill). The top-loaders used between 40 and 52 gallons in the regular cycle. The front-loader used 24 gallons. About half the machines used several gallons more water in the permanent-press cycle.

7 Energy efficiency. Based on gallons of hot water used in hot/cold cycles. Most of the energy cost of a washing machine comes from heating hot water; the motor uses only a few dollars' worth of electricity a year. The typical top-loader doing six loads a week would need $112 worth of electrically heated water a year or $37 worth of gas-heated water, at average utility rates. The front-loader would use only $42 in electricity or $14 in gas for the same amount of laundry. The use of a warm wash would halve these figures. If you do all your washes in cold water, you can consider the machines equally energy-efficient.

8 Handling unbalanced loads. How well a machine could cope with off-kilter loads. Some stop too soon; the best in these tests could complete the spin cycle with a severely unbalanced load. Average machines could handle a moderately unbalanced load with only a little protest. The tub in the most sensitive models dragged or banged, preventing the spin cycle from working properly, or it shut off altogether.

9 Extraction. How well the spin cycle removes water. The wetter the finished wash, the longer it will take to dry.

10 Linting. Some filter systems worked better than others, but even the best ones deposited some lint from diapers and white towels on dark socks. A judgment that matters most to those who don't sort their laundry.

Better ←————→ Worse

1 Brand and model	2 Price	3 Dimensions (HxWxD), inches	4 Lid opens	5 Load capacity	6 Water efficiency	7 Energy efficiency	8 Unbalanced loads	9 Extraction	10 Linting	11 Noise	12 Service access	Advantages	Disadvantages	Comments
Top-loaders														
Sears Kenmore 29801	$420	43¼(53½)x27x25¾	Left	◉	◓	◓	○	◉	○	◉	◉	A,B,C,J	a,b,m	A,I
KitchenAid KAWE 550	420	42½(51½)x27x25¾	Back	◉	◓	◓	○	◉	◓	◉	◉	A,I	a,c	A,I,J
Whirlpool LA5558XS	377	42½(51½)x27x25¾	Back	◉	◓	◓	○	◉	○	◉	◉	A,J	a,c,m,n	A,I
Maytag A9700	610	43¾(51)x25½x27	Back	◓	◉	◉	◓	◉	◓	◉	◒	A,B,I,K	c,d,e,h	B,D,I
Gibson WA27M4	385	43¾(51)x27x27	Back	◓	◓	○	◉	◓	○	○	◒	D,F,G,J,K	h	—
Kelvinator AW700G	363	43½(51)x27x27	Back	◓	◓	○	◉	◓	○	○	◒	D,F,G,J	h,n	—
Magic Chef W20H3	440	44(54)x27x27	Back	◓	◓	○	◒	◓	◓	◓	◒	B,E,J,K	f	F
Admiral AW20K-3	422	44(54)x27x27	Back	◓	◓	○	◒	◓	◓	◓	◒	B,E,J,K	f	F
Montgomery Ward 6530	420	44(54)x27x27	Back	◓	◓	○	◒	◉	◓	◓	◒	E,J,K	f,j	F,J
White-Westinghouse LA500M	375	43½(51)x27x27	Back	○	○	◓	◒	◉	○	◓	◒	D,F,G,J	e,h,n	—
Frigidaire WCDL	378	44(54½)x27x27	Left	○	◓	◒	◒	◉	○	◓	◒	D,F,G,J	b,n	—
Amana LW2303	427	42(50¼)x25¾x28	Back	○	○	◓	○	◓	○	○	◒	A,H	h,k,n	D,J
Speed Queen NA4521	457	42¼(50¼)x25¾x28	Back	○	○	◓	○	◓	○	○	◒	A,H	h,k,n	D,J
General Electric WWA8324G	392	43½(50½)x27x25	Back	○	◓	◓	◒	◉	◓	○	◒	I	f,g,n,o	D,G,J
RCA WRW3705K	385	43½(50½)x27x25	Back	○	◓	◓	◒	◉	○	○	◒	I	f,g,l	D,G
Hotpoint WLW3700B	375	42½(50½)x27x25	Back	○	◓	◓	◓	◒	○	○	◒	I	f,g,l	D,G
Front-loader														
White-Westinghouse LT250L	568	34¾x27x27(39¼)	Down	◓	●	◉	◉	◓	◉	◉	◒	A,B,D	i,j,m,n	C,E,H

⑪ Noise. A judgment that matters if the machine is near the living area or next to a wall that adjoins someone else's.

⑫ Service access. Easiest in models whose wraparound cabinet can be removed in one piece. Next best were models with a removable front panel. Worst were machines whose innards are approachable only from the rear.

Performance Notes

Except as noted, all were judged average in sand disposal and handling of permanent-press items for line drying.

Features in Common

All have: ● At least three cycles. ● Timer dial for cycle selection. ● Water-level control. *Except as noted, all have:* ● Lid safety switch that stops spin action only. ● Self-cleaning lint filter system, bleach and fabric-softener dispensers. ● Comprehensive operating instructions on lid. ● Top, lid, and cabinet with plastic or painted finish. ● Tub with porcelain enamel finish.

Key to Advantages

A—Better than average in handling permanent-press items for line drying.
B—Soak/prewash cycle.
C—Electronic water-temperature control; if selected, blends hot and cold water to preset temperature.

D—Severely unbalanced load caused only slight vibration.
E—Safety switch stopped spinning tub more quickly than most.
F—Lid locks during spin.
G—Plastic tub.
H—Stainless-steel tub.
I—Top and lid finish of porcelain enamel.
J—Has self-adjusting legs at rear.
K—Easy-to-clean rinse-agent cup.

Key to Disadvantages

a—Water-level control somewhat hard to turn.
b—Lid does not open fully, hampering access from left.
c—No lip on top to contain minor spills.
d—Tub took longer than others to stop when lid was opened during spin.
e—No instructions on lid.
f—Moderately unbalanced load caused knocking and reduced water extraction.
g—Tub brake clanks at end of spin.
h—Slightly poorer than average in sand disposal.
i—Door hard to open and close.
j—Timer control somewhat difficult to adjust.
k—Moderately unbalanced load stopped spin.
l—Changing speed setting while machine is

running may damage drive mechanism.
m—No bleach dispenser.
n—No fabric-softener dispenser.
o—Lint filter not self-cleaning; must be removed for access to tub.

Key to Comments

A—Severely unbalanced load caused loud banging until tub picked up sufficient speed.
B—Severely unbalanced load stopped spin.
C—Lid locks during spin (and most other parts of cycle), but lock can be defeated by turning timer to off.
D—Lid safety switch stops agitation and spin.
E—Tub light.
F—Tub has plastic finish.
G—Limited instructions on lid.
H—Model intended for under-counter or stacked dryer installation; controls are on front.
I—Agitator has slots under vanes. Some consumers have complained it snags laundry, but that did not happen in the tests.
J—Although this model has been discontinued and is no longer available, the information has been retained to permit comparisons.

Index